Economic Benefits
of Improved Water Quality

Other Titles in This Series

Water and Western Energy: Impacts, Issues, and Choices, Steven C. Ballard, Michael D. DeVine, and Associates

Water and Agriculture in the Western U.S.: Conservation, Reallocation, and Markets, edited by Gary D. Weatherford

Also of Interest

Western Water Resources: Coming Problems and the Policy Alternatives, A Symposium Sponsored by the Federal Reserve Bank of Kansas City

Scientific, Technological and Institutional Aspects of Water Resource Policy, edited by Yacov Y. Haimes

Water and Energy in Colorado's Future: The Impacts of Energy Development on Water Use in 1985 and 2000, Colorado Energy Research Institute

Water Needs for the Future: Political, Economic, Legal, and Technological Issues in a National and International Framework, edited by Ved P. Nanda

Public Representation in Environmental Policymaking: The Case of Water Quality Management, Sheldon Kamieniecki

**Energy Futures, Human Values, and Lifestyles: A New Look at the Energy Crisis*, Richard C. Carlson, Willis W. Harman, Peter Schwartz, and Associates

**Renewable Natural Resources: A Management Handbook for the 1980s*, edited by Dennis L. Little, Robert E. Dils, and John Gray

Energy, Economics, and the Environment: Conflicting Views of an Essential Relationship, edited by Herman E. Daly and Alvaro F. Umaña

**The Economics of Environmental and Natural Resources Policy*, edited by J. A. Butlin

*Available in hardcover and paperback.

Studies in Water Policy and Management
Charles W. Howe, General Editor

*Economic Benefits of Improved Water Quality:
Public Perceptions of Option and Preservation Values*
Douglas A. Greenley, Richard G. Walsh, and Robert A. Young

Until recently, there has been general agreement that improvement and preservation of water quality, though costly, provided economic and social benefits that outweighed the expense. Now, however, some observers are beginning to question whether the costs of the 1972 Water Pollution Control Act may actually exceed those benefits. This book provides answers to some of the questions that have been raised.

The authors give measures of several important nonmarket benefits of improved water quality in Colorado's South Platte River Basin and empirically test and confirm the Weisbrod and Krutilla proposals that the general public may be willing to pay for preservation of environmental amenities and that option value and other preservation values must be added to recreation-use values to give an accurate picture of the social benefits of environmental preservation and restoration. Their findings include the fact that even those who do not expect to use the river basin for recreation are willing to pay for the maintenance of a natural ecosystem and to bequest clean water to future generations. The authors also arrive at average amounts households are willing to pay for improved water quality to enhance enjoyment of water-based recreation activities. They suggest that, without such information, it is highly unlikely that sufficient resources will be allocated for the preservation of unique environments and for the improvement of those being degraded.

Douglas A. Greenley is an associate professor of economics at Moorhead State University in Minnesota. *Richard G. Walsh*, a former Intergovernmental Exchange Scholar with the Environmental Protection Agency, is professor of economics at Colorado State University and specializes in studies of recreational demand. *Robert A. Young*, former research associate with Resources for the Future, is professor of economics at Colorado State University. He is the author of *The Economic Value of Water* and numerous other publications on water resources and environmental quality.

Economic Benefits of Improved Water Quality
Public Perceptions of Option and Preservation Values

Douglas A. Greenley, Richard G. Walsh, and Robert A. Young

Studies in Water Policy and Management, No. 3

Westview Press / Boulder, Colorado

ST. PHILIP'S COLLEGE LIBRARY

Studies in Water Policy and Management

All rights reserved. No part of this publication may be reproduced or transmitted in any form or by any means, electronic or mechanical, including photocopy, recording, or any information storage and retrieval system, without permission in writing from the publisher.

Copyright © 1982 by Westview Press, Inc.

Published in 1982 in the United States of America by
    Westview Press, Inc.
    5500 Central Avenue
    Boulder, Colorado 80301
    Frederick A. Praeger, President and Publisher

Library of Congress Catalog Card Number: 82-50434
ISBN 0-86531-414-4

Composition for this book was provided by the authors
Printed and bound in the United States of America

Contents

List of Tables xi
List of Figures. xv
Foreword, *Charles W. Howe*. xvii
Acknowledgments. xix

1 INTRODUCTION AND SUMMARY. 1

   Background and Scope of Study 1
   Objective 5
   Methodology 6
   Summary of Results. 6
   Outline 7
   Notes 7

2 PREVIOUS RESEARCH ON RECREATION AND PRESERVATION
   BENEFITS OF WATER QUALITY 9

   Recreation Benefits from Improved Water Quality 9
      Travel Cost Approach. 11
      Contingent Value Approach 12
      Unit Day Value. 13
      Recreation Benefit Studies. 15
   Preservation Benefits from Improved Water Quality 24
      Option Value. 25
      Existence Value 26
      Bequest Value 27
      Preservation Benefit Studies. 28
   Notes 30

3 THE SOUTH PLATTE RIVER BASIN, COLORADO. 35

   Selection Criteria. 35
   Environmental and Economic Characteristics. 37
   Water Quality 38
   Consequences of Mining in Colorado. 39
   Notes 44

viii

4	DATA COLLECTION PROCEDURES.	45
	Sample Selection.	45
	Characteristics of Sample and Population.	46
	Establishing Contact with Respondents	49
	Personal Interviews	50
	Simulated Market Payment Vehicles	51
	Iterative Contingent Value Approach	59
	Notes	62
5	OPTION, PRESERVATION, AND RECREATION BENEFIT ANALYSIS . . .	63
	Calculation of River Basin Benefits	63
	Option Values	64
	Recreation Use Values	65
	Existence and Bequest Values.	70
	Total Annual Benefits Per Household	74
	Annual Benefits and Present Value of Future Benefits. . . .	75
	Method of Payment	77
	Level of Water Quality.	78
	Effects of Delayed Water Quality Improvement.	79
	River Basin Versus State Values	84
	Who Should Bear the Cost of Improved Water Quality?	87
	Environmental Awareness	87
	Inter-City Comparisons.	90
	Notes	91
6	SOCIOECONOMIC RELATIONSHIPS	93
	Household Income.	93
	Sex of Respondent	97
	Employment.	97
	Education	101
	Former Residence.	101
	Reasons for Moving.	104
	Permanence of Residence	104
	Age of Respondent	107
	Size of Household	107
	Recreation Use.	110
	Notes	113
7	SUMMARY AND CONCLUSIONS	115
	BIBLIOGRAPHY.	123
	APPENDIX A: CONCEPT OF OPTION DEMAND	131
	Notes	136
	APPENDIX B: QUESTIONNAIRE.	139
	Facsimile of Introductory Letter.	140
	Water Quality Opinion Survey.	141

APPENDIX C: REGRESSION ANALYSIS. 145

Specification of the Regression Models. 145
  Variables Used in Analysis of Relationship Between
  Willingness to Pay and Socioeconomic Characteristics. . . 146
Interpretation of Dummy Variables 147
Statistical Reliability of the Socioeconomic Regressions. . 148
Notes 150

INDEX 159

Tables

1.1 Estimated Stream and Shoreline Pollution in Major River Basins, United States, 1971. 2
3.1 Production of Metals in Colorado, 1973-75. 40
3.2 Comparative Drinking Water and Biological Standards for Trace Elements 43
4.1 Sample Response in Denver and Fort Collins, Colorado, 1976 46
4.2 Comparison of Population and Sample Demographic Profiles for Denver and Fort Collins, Colorado, 1976. 47
4.3 Heavy Metal Pollution at the Three Photograph Sites, South Platte River Basin, Colorado, 1973 53
4.4 Total Number of Residents Interviewed Compared to the Number Answering Water Quality Valuation Questions and the Number Willing to Pay in the South Platte River Basin, Colorado, 1976 55
4.5 Five Percent and One-Quarter Percent Incremental Sales Tax Value Estimates for Colorado Residents by Income and Family Size, 1975. 57
5.1 Resident Household Mean Willingness to Pay Additional Sales Taxes and Water Service Assessments for Water Quality in the South Platte River Basin, Colorado (1981 dollars). . . 66
5.2 Resident Recreation User and Nonuser Willingness to Pay Additional Water Service Assessments and Sales Taxes to Improve Water Quality for Existence and Bequest Benefits in the South Platte River Basin, Colorado (mean values in 1981 dollars) 72
5.3 Annual and Present Value of Water Quality in the South Platte River Basin, Colorado (1981 dollars). 76
5.4 Effect of Level of Pollution Abatement on Resident Household Willingness to Pay Additional Water Service Assessments and Sales Taxes to Improve Water Quality for Recreation Use in the South Platte River Basin, Colorado (mean values in 1981 dollars). 80
5.5 Effect of Delay from 1983 to 2000 on Resident Household Willingness to Pay Additional Water Service Assessments and Sales Taxes to Improve Water Quality for Recreation in the South Platte River Basin, Colorado (mean values

	in 1981 dollars)	82
5.6	South Platte River Basin (SPRB) Resident Household Willingness to Pay Additional Water Service Assessments and Sales Taxes to Improve Water Quality for Recreation Throughout Colorado (mean values in 1981 dollars).	85
5.7	Resident Opinions as to Who Should Pay for Water Quality in the South Platte River Basin, Colorado, 1976.	88
5.8	Environmental Awareness of Residents in the South Platte River Basin, Colorado, 1976.	89
6.1	Regression Coefficients of Significant Socioeconomic Variables, Denver and Fort Collins, Colorado, 1976	94
6.2	Household Income and Willingness to Pay Additional Sales Tax for Improved (C-A) Water Quality, Denver, Fort Collins, and South Platte River Basin, Colorado, 1976. . .	95
6.3	Marginal Effect of a Change of Income on Willingness to Pay Additional Sales Tax for Improved Water Quality, at Various Age Levels, Denver Metropolitan Area, Colorado, 1976	96
6.4	Sex of Respondent and Willingness to Pay Additional Sales Tax for Improved (C-A) Water Quality, Denver, Fort Collins, and South Platte River Basin, Colorado, 1976	98
6.5	Type of Employer Related to Willingness to Pay Additional Sales Tax for Improved (C-A) Water Quality, Denver, Fort Collins, and South Platte River Basin, Colorado, 1976. . .	99
6.6	Occupation and Willingness to Pay Additional Sales Tax for Improved (C-A) Water Quality, Denver, Fort Collins, and South Platte River Basin, Colorado, 1976	100
6.7	Education and Willingness to Pay Additional Sales Tax for Improved (C-A) Water Quality, Denver, Fort Collins, and South Platte River Basin, Colorado, 1976	102
6.8	Size of Place of Previous Residence and Willingness to Pay for Improved (C-A) Water Quality, Denver, Fort Collins, and South Platte River Basin, Colorado, 1976	103
6.9	Reason for Moving to Colorado and Willingness to Pay Additional Sales Tax for Improved (C-A) Water Quality, Denver, Fort Collins, and South Platte River Basin, Colorado, 1976	105
6.10	Permanence of Residence and Willingness to Pay Additional Sales Tax for Improved (C-A) Water Quality, Denver, Fort Collins, and South Platte River Basin, Colorado, 1976. . .	106
6.11	Age and Willingness to Pay Additional Sales Tax for Improved (C-A) Water Quality, Denver, Fort Collins, and South Platte River Basin, Colorado, 1976	108
6.12	Size of Household and Willingness to Pay Additional Sales Tax for Improved (C-A) Water Quality, Denver, Fort Collins, and South Platte River Basin, Colorado, 1976. . .	109
6.13	Survey Respondents Reported Annual Water-Based Recreation Activity Days in the South Platte River Basin and Willingness to Pay Additional Sales Tax for Improved (C-A) Water Quality, Denver, Fort Collins, and South Platte River Basin, Colorado, 1976.	111
6.14	Survey Respondents Reported Annual Water-Based Recreation Activity Days in the United States and Willingness to	

xiii

	Pay Additional Sales Tax for Improved (C-A) Water Quality, Denver, Fort Collins, and South Platte River Basin, Colorado, 1976. 112
C-1	Stepwise Multiple Regression of Resident Households' Willingness to Pay Additional Sales Taxes to Improve Water Quality for Recreation Use, Denver, Colorado, 1976 151
C-2	Stepwise Multiple Regression of Resident Households' Willingness to Pay Higher Water Service Prices to Improve Water Quality for Recreation Use, Denver, Colorado, 1976 152
C-3	Stepwise Multiple Regression of Resident Households' Willingness to Pay Higher Water Service Prices to Improve Water Quality for Recreation Use, Fort Collins, Colorado, 1976 153
C-4	Stepwise Multiple Regression of Resident Households' Willingness to Pay Additional Sales Taxes for Option Value from Preserved Water Quality, Denver, Colorado, 1976 . . . 154
C-5	Stepwise Multiple Regression of Resident Households' Willingness to Pay Additional Sales Taxes for Option Value from Preserved Water Quality, Fort Collins, Colorado, 1976 155
C-6	Stepwise Multiple Regression of Resident Households' Willingness to Pay Higher Water Service Prices for Option Value from Preserved Water Quality, Denver, 1976 156
C-7	Stepwise Multiple Regression of Resident Households' Willingness to Pay Higher Water Service Prices for Option Value from Preserved Water Quality, Fort Collins, Colorado, 1976 157

Figures

1.1 Aggregate Benefit and Cost Functions for Water Quality . . . 4
2.1 Shift in Demand Curve with Improved Water Quality at a
    Recreation Site. 10
3.1 The South Platte River Basin, Colorado 36
4.1 Photographs of Three Stream Sites, South Platte River
    Basin, Colorado, 1976. 52
4.2 Price Compensating Variation of Consumer Surplus 62

Foreword

This volume dealing with the public's perception and valuation of improvements in water quality is part of Westview's Studies in Water Policy and Management, which address contemporary water problems, expanding and clarifying the range of policy alternatives available for solutions. The series emphasizes the economic, legal, political, and administrative dimensions of water systems and is intended not only for scholars but also for technical and political decision makers.

Water quality is a nonmarket good that links all water users. It is important to know the benefits of improving water quality because improvements involve either costs of treatment and process change or reductions in economic activities. The benefits of improving water quality include recreational values of active users, the willingness to pay for preserving the option of higher water quality by potential recreational users, and a general willingness to pay for preserving natural ecosystems for current and future generations. Through survey techniques, this study shows that these values are quite substantial and should be added to the more common measures of reductions in agricultural and municipal damages.

Charles W. Howe
General Editor

Charles W. Howe, editor of Westview's Studies in Water Policy and Management, is a professor of economics at the University of Colorado. He has served as director of the Water Resources Program, Resources for the Future, Inc., and on the Board of Editors of Land Economics *and the American Geophysical Union Water Monograph Series. Dr. Howe's previous books include* Natural Resource Economics: Issues, Analysis, Policy *and* Managing Renewable Natural Resources in Developing Countries *(Westview, 1982).*

Acknowledgments

We wish to express our thanks to Dr. Anthony Fisher, University of California, Berkeley; Dr. John Krutilla, Resources for the Future, Washington, D.C.; Dr. Alan Randall, University of Kentucky, Dr. Phillip Meyer, Fisheries and Marine Service, British Columbia, Canada; Dr. Lawrence Leuzzi, University of Hawaii, and Jonathan Scherschlight, Water Quality Control Division, Colorado Department of Health, for their assistance in development of the conceptual framework and questionnaire used in this study. A number of colleagues at Colorado State University have contributed to this work. Dr. Anthony Prato helped prepare a summary of the literature on option value and Dr. John McKean provided advice regarding the statistical analysis. The authors are, of course, responsible for any shortcomings which may remain.

The research was funded, in part, by the U.S. Environmental Protection Agency, which reviewed and approved the results for publication. However the study does not necessarily reflect the views and policies of the Agency. The advice of Fred Abel, Dennis Tihansky, Thomas Waddell, and Donald Gillette, EPA Project Officers, has been greatly appreciated. We also wish to acknowledge the support by the Colorado State University Experiment Station of the sample survey and a portion of the principal investigators' salaries.

The authors also thank Mr. Austin Buckman for interviewing assistance and Ms. Waneta Enyeart Boyce for typing the manuscript.

D.A.G.
R.G.W.
R.A.Y.

1
Introduction and Summary

BACKGROUND AND SCOPE OF STUDY

The Water Pollution Control Act amendments of 1972 (P.L. 92-500) provided the necessary legal framework for an ambitious effort aimed at extensive improvement of water quality in lakes and streams throughout the United States. As specified in the Act, the national goal was no less than "water quality which provides for the protection and propagation of fish, shellfish, and wildlife and provides for recreation in and on the water . . ." with the eventual elimination of all effluent discharge. The law called for all industrial dischargers to have installed the best practical treatment methods by 1977 and the best available treatment technology by 1983. Municipal treatment systems were to have installed secondary treatment by 1977 and the best practicable waste water treatment technology by 1983. Non-point sources of pollution such as erosion and runoff from agricultural and urban areas were required to develop and implement area-wide waste treatment management plans.

The Clean Water Act of 1977 (P.L. 95-217) reaffirmed the national goal of fishable and swimmable water quality by 1983 and established a new goal of eliminating the discharge of pollutants into navigable waters by 1985. Amendments passed in 1977 modified the best available treatment requirements and extended the 1983 deadline for various industries. This is not surprising, since these goals are expensive and perhaps impossible to attain. Competing national goals of expanded mineral and energy production and inflation reduction may conflict with the attempt to achieve water quality. Moreover, the benefits to be achieved by the Federal goals may not be sufficient to justify the necessary public and private expenditures. The Environmental Protection Agency has come under increasing pressure to assess its water quality improvement programs in terms of their benefits and costs.

There was widespread agreement that the problem of water pollution in the United States had reached unacceptable levels when the 1972 water quality legislation was passed. The Environmental Protection Agency (EPA) estimated that 29 percent of all U.S. river and shoreline miles were polluted in 1971. Table 1.1 shows the amount of polluted water reported in the nine major river basins of the

Table 1.1. Estimated Stream and Shoreline Pollution in Major River Basins, United States, 1971.

Major River Basin	Total Stream and Shoreline (Miles)	Polluted Stream and Shoreline (Miles)	Proportion Polluted (Percent)
Ohio	28,992	24,031	83
Southeast	11,726	4,490	38
Great Lakes	21,374	8,771	41
Northeast	32,431	5,823	18
Middle Atlantic	31,914	5,627	18
California	28,277	8,429	30
Gulf	64,719	11,604	18
Missouri	10,448	1,839	18
Columbia	30,443	5,685	19
United States	260,324	76,299	29

Source: Environmental Protection Agency, The Cost of Clean Water, (Washington, D.C., 1972).

U.S. Nearly all of the waterways in the Ohio River Basin were polluted, and waterways in the Great Lakes and Southeast drainage areas exceeded the national average pollution level by one-third. California waterways were nearly identical in quality to the national average with 30 percent polluted.

Measuring the level of water quality is a very complex question. Pollution is always a matter of degree, rather than yes or no, and defining pollution is closely related to the use to which water may be put. The estimates were based on Federal-State water quality standards which vary from place to place, depending on the local use of water for drinking, swimming, fishing, industrial waste discharge, and other uses. The estimates were corrected for natural pollution levels. In the EPA study, the calculation of percentage of water polluted was supplemented using a prevalance-duration-intensity index showing the degree of pollution and how long during the year the waterway was in violation of the standards.

A specialized measure of water quality for Colorado showed substantial pollution levels. The Colorado Division of Wildlife reported that 8,700 miles or 70 percent of the 12,500 miles of river in the state were capable of sustaining a trout population in 1970.[1] An estimated 2,640 miles or 21 percent of the rivers in the state were polluted; 917 miles of river or 7 percent were dewatered by irrigation and power users; and 253 miles or 2 percent were inundated by reservoir construction.

In the past, most Western communities and state governments welcomed and encouraged the clean water program as a source of new income and general economic growth. Recently, some observers have begun to question whether the social benefits of water pollution control exceed social costs. The people involved are interested in what can be learned from current experience to help formulate sound water quality policies for the future. The purpose of this study is to begin to provide answers to some of the questions which have been raised. Procedures are developed to measure several important non-market benefits from improved water quality in the South Platte River Basin, Colorado.

A complete economic analysis of a proposed policy to improve water quality would measure: (1) the benefits of pollution control, (2) the costs of reducing or removing waste discharge, (3) the costs of monitoring and enforcing regulations, and (4) where chronic unemployment exists, the indirect or secondary benefits and costs.[2] A recommended program on grounds of economic efficiency would be one for which the incremental benefits exceed the incremental costs. Bradford[3] proposed the use of aggregate benefit or bid functions as shown in Figure 1.1 which are derived from the vertical summation of benefits of the affected population. When total costs are known, an optimum level of water quality can be determined where the difference between total benefits and total costs is maximized or at the point where the slopes of the two curves are equal. Alternatively, the first derivatives of total benefit and cost curves yields the marginal benefit and cost curves shown in Figure 1.1. An efficient allocation of resources occurs at the point where the marginal benefit and cost curves intersect.

The total costs of pollution control programs have been estimated by the National Commission on Water Quality[4] and others. However, there are very few investigations of the aggregate social benefits of improved water quality. Early studies focused on the recreation-associated benefits of water quality improvement and ignored any economic assessment of the preservation value to the general population.

This book provides an empirical test and confirmation of the Weisbrod[5] and Krutilla[6] proposals that: the general public may be willing to pay for the preservation of environmental amenities; option value and other preservation values represent important social benefits; and should be added to recreation use values to determine the total benefit of environmental amenities to society. In the absence of information on preservation benefits to all of the people, insufficient resources would be allocated by society to the preservation of unique environments such as pristine mountain streams where mineral and energy development may irreversibly degrade water quality.

Measuring the economic value of the preservation benefits of environmental quality has proven to be a difficult aspect of an already complex problem. In addition to the usual difficulties encountered in measuring the value to society of recreation use where market transactions are absent, the benefits from pollution abatement also include significant option, existence, and bequest values. The primary contribution of this analysis is to empirically test the importance of these latter values relative to the conventional recreation benefits. Measurement of these nonmarket and rather abstract values such as option, existence, and bequest demand requires careful development of a methodology which allows the assessment of the value of water pollution abatement to members of the appropriate population.

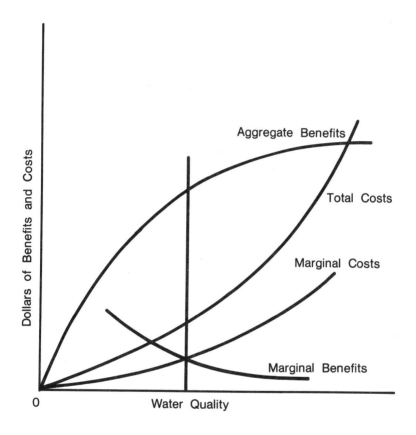

Figure 1.1. Aggregate Benefit and Cost Functions for Water Quality.

Krutilla noted several possible instances of willingness to pay for attributes of environmental quality which are distinct from the direct or immediate benefits to users of a natural resource. These previously unrecognized benefits of environmental quality were termed preservation benefits by Krutilla. Since this book develops and implements a procedure for measuring preservation benefits, it is appropriate to delineate just what is being measured.

Option value has been the subject of considerable controversy among economists. The debate culminated in the development of a definition and model by Henry[7] which is tested in this study. Option value is taken to mean the premium that individuals would be willing to pay to preserve irreplaceable environmental resources now, in order that a more informed choice can be made at some future date, when the necessary information affecting their decision of whether or not to preserve the environment is available. This construct differs from earlier notions of option value. The previous concept of option value stated that a risk averse individual would be willing to pay to preserve his option to make recreation use of a facility in the future when there is a threat of irreversible damage to the natural environment. Existence value is simply the willingness to pay for the knowledge that the natural environment is preserved. Bequest value, which seems closely related to existence value, is the worth to present generations from preserving the environment for future generations.

OBJECTIVE

The purpose of this study was to develop and apply a procedure for analyzing the benefits of improved water quality to both recreation users of the resource and the general population. The South Platte River Basin located in northeastern Colorado was selected as the study area.

Specific objectives were to:

1. Develop a conceptual framework and empirically test its application in the measurement of benefits of water quality improvement. The benefits measured include:

    a. Consumer surplus from enhanced enjoyment of water-based recreation activities;
    b. Option value of assured choice of recreation use in the future through avoidance of irrevocable pollution by mineral and energy development, and
    c. Existence and bequest values of the resident population.

2. Identify the relationship between these values and the quality of water available as measured on a three-point scale from low to high.
3. Test statistically the relationship between the expressed values of improved water quality and socioeconomic variables including income, degree of urbanization, education, age, occupation, amount of water-based recreation, and family size.

METHODOLOGY

The economic values reported in this study are based on the contingent value approach recently recommended by the U.S. Water Resources Council[8] as suitable for studies of recreation and environmental quality. The approach relies on the stated intentions of individuals to pay with variations in water quality depicted in color photographs. The iterative bidding technique was used to assure that maximum points of indifference between having the stated amount of income or level water quality were reported.

Two hundred and two randomly selected resident households in Denver and Fort Collins, Colorado, were interviewed in their homes during the summer of 1976. Respondents were shown color photos of three representative stream sites in the South Platte River Basin. Heavy metal content served as an index of water quality ranging from low to high.

Given the three varying levels of water quality, residents were asked by how much they would be willing to increase payment of annual sales taxes and monthly water service fees to improve basin lakes and streams for enhanced recreational opportunities. Respondents were also queried on their willingness to pay for option, existence, and bequest benefits from improved water quality.

SUMMARY OF RESULTS

Water quality improvement benefits were originally estimated in 1976 and updated to 1981. Benefit values were conservatively estimated to have inflated by 41 percent over this five year period, based on changes in the GNP Implicit Price Deflator.

The estimated average annual total benefit from water quality improvement for a household currently engaging in water-based recreation and also possessing option, existence, and bequest demands was $170 in increased sales taxes. The estimate was made under the assumption that water quality would be improved to a high level by 1983.

The largest single value associated with water quality improvement was for enhanced current water-based recreation activities. The reported values amounted to $80 per resident household as measured by willingness to pay increased sales taxes and $26 in higher water service fees. Option value estimates were equivalent to about 40 percent of resident recreation use benefits, averaging $32 in sales tax revenues and $11 in water service charges. Existence and bequest values were measured using a subsample of nonrecreationists who reported a zero probability of future water-based recreation activity in the basin. The 19.2 percent of such nonusers surveyed reported existence values of $35 and $9 per household annually through increased sales taxes and water service charges, respectively. Bequest values for nonrecreationists were estimated at $24 in sales tax revenues and $8 in higher water service charges. Existence and bequest value estimates for the subsample of nonrecreationists were extended to the general population.

The total present value from improved water quality was estimated as $1.3 billion in additional sales tax. This included recreation use benefits of $450 million, option value of $254 million, existence value of $347 million, and bequest value of $236 million.

Socioeconomic variables such as income and education were regressed on recreational use and option value estimates. The results suggest some important explanatory relationships. Variables such as income, male respondents, educational attainment, and residents migrating from smaller cities to Denver were all positively associated with willingness to pay for water quality improvement. Government employees valued enhanced recreational use more than employees in the private sector while those in professional occupations, business owners and managers valued it less than other professions. Housewives were also willing to pay more for water quality improvement for recreational use in Denver. Retired residents were willing to pay less than those employed.

OUTLINE

This book is organized into seven chapters and an appendix. To provide perspective, Chapter 2 contains a brief review of past attempts to empirically estimate recreation and preservation benefits from water quality improvement. Chapter 3 includes an overview of existing environmental and socioeconomic characteristics of the South Platte River Basin in Colorado. Also included is a brief review of the effects of imminent mineral and energy development. A discussion of the survey methodology employed in the study is provided in Chapter 4. Option, preservation, and recreation benefits are analyzed in Chapter 5. Chapter 6 shows the results of a regression of socioeconomic characteristics of basin residents on willingness to pay for recreation use and option value. Also shown are cross tabulations of the more important socioeconomic variables with willingness to pay. A brief summary and conclusions of the study are provided in Chapter 7. The appendix includes a review of the theory of option value, the questionnaire, and the regression analysis.

NOTES

1. Thomas Lynch, "Population, Pollution Limit Fishing," Coloradoan, (Fort Collins, 1970).
2. Allen V. Kneese and W. D. Schulze, Pollution, Prices, and Public Policy, (Washington: Brookings Institution, 1975).
3. David F. Bradford, "Benefit-Cost Analysis and Demand Curves for Public Goods," Kyklos, 23 (1970), 775-91.
4. National Commission on Water Quality, Staff Report, (Washington, D.C., 1976).
5. Burton A. Weisbrod, "Collective-Consumption Services of Individualized-Consumption Goods," Quarterly Journal of Economics, 78 (August 1964), 471-77.
6. John V. Krutilla, "Conservation Reconsidered," American Economic Review, 57 (September 1967), 777-86.

7. Claude Henry, "Option Values in the Economics of Irreplaceable Assets," *The Review of Economic Studies: Symposium on the Economics of Exhaustible Resources*, (1974), 89-104.

8. U.S. Water Resources Council, "Procedures for Evaluation of National Economic Development (NED) Benefits and Costs in Water Resources Planning," *Federal Register*, 44:242 (December 14, 1979), 72, 950-65.

2
Previous Research on Recreation and Preservation Benefits of Water Quality

Pollution of a river or lake, to the extent that it diminishes the satisfaction of some individuals, is viewed as a damage while improved water quality is considered a benefit. The problem is especially difficult because an improvement in water quality which reduces this external cost of pollution is a nonmarket product, since it is nonexclusive, and a public good since it is inexhaustible at least over a substantial range. Economic analyses of nonmarket and public goods are in an experimental stage. Problems include conceptual ambiguities and incomplete data from which to identify damage relations. Thus, it is not surprising that efforts to determine the magnitude of these benefits provide only tentative estimates at best. Before discussing the results of our study it will help provide perspective to review the past research on the economic significance of recreation and preservation benefits of water quality.

RECREATION BENEFITS FROM IMPROVED WATER QUALITY

It will be helpful first to review the conceptual basis for defining and measuring the benefits of improved water quality for recreation use.[1] Figure 2.1 shows two demand curves for a recreation site with and without water pollution. Each demand curve is plotted from a demand function which relates quantity of recreation services demanded to costs of access and participation, holding constant the effects of water quality, congestion, income, the prices and availability of substitutes, and other important determinants of demand for sites in a region. At an average price or direct cost of A, a total of Q_1 recreation days will be consumed without improved water quality. The value of the recreation site, given the initial polluted water, is consumer surplus measured as the area ABC.

With improved water quality, the demand curve shifts to the right, and Q_2 will be demanded at a price of A. The net economic benefit of improved water quality is the increase in willingness to pay measured as the area between the two demand curves, BCED. This can be divided into two categories. Area BCFD is the benefit of improved water quality to those who were using the site before water quality improved. Area CEF is the benefit of improved water quality associated with increased use rates by the original users and other

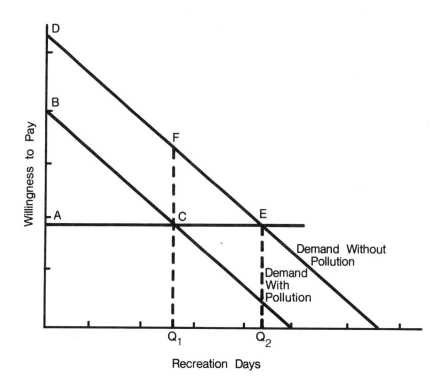

Figure 2.1. Shift in Demand Curve with Improved Water Quality at a Recreation Site.

users who are attracted to the site by the improved water quality. Under conditions of excess demand for some sites in the region, this substitution of a site with improved water quality may reduce congestion at other sites. The area between demand curves with and without this reduced congestion at other sites (not shown) is the benefit to the remaining users who did not substitute. Since this is a consequence of improved water quality, the benefit of reduced congestion at other sites should be added to the benefit of improved water quality at a single site.[2]

Three methods have been recommended by the U.S. Water Resources Council[3] as providing acceptable economic measures of the recreation benefits of environmental change. The interagency committee established uniform procedures for application of the travel cost, contingent value, and unit day methods. The travel cost approach to the estimation of demand for recreation use has been preferred by most economists, since it is based on observed market behavior of a cross-section of users in response to out-of-pocket and time cost of travel and other variables such as water quality. The contingent value approach relies on the stated intentions of a cross section of the affected population to pay for an environmental amenity contingent on hypothetical changes depicted in color photos. The value reported is assumed to correspond to the point of indifference between having that amount of income or the environmental amenity. This study is an application of the contingent value approach to estimate the recreation use benefits and preservation values of water quality to the resident population of the South Platte River Basin. Unit day values are selected from a range of values approved by the Council and other agencies. Initially based on a survey of entrance fees at private recreation areas in 1960, unit day values have been adjusted for changes in the Consumer Price Index to the present.

Travel Cost Approach[4]

The travel cost approach to estimation of the demand effect of changes in water quality is based on observed behavior of a sample of users responding to changes in out-of-pocket and time costs of traveling to: (1) a single recreation site before and after water quality is improved, or (2) a cross section of recreation sites with varying water quality. When a cross section of recreation sites is included in a regional demand equation, the regression coefficient for water quality provides a statistical estimate of the shift in demand associated with changes in water quality, all other variables in the equation remaining constant. The dependent variable is annual trips, and the independent variables are: travel cost; travel time; price and availability of substitutes; characteristics of the site including water quality, congestion, and surface area; income and other socioeconomic variables.

In the first stage, a statistical estimate of this relationship is derived from a multiple regression with number of trips per individual or per 1,000 population in concentric zones or counties as the dependent variable. The observed number of trips by all individuals or from all distance zones reflects demand at current travel costs. This represents one point on the second stage demand curve, the number of visits with no price increase. The second stage of the travel cost approach consists of calculating total recreation use at each of a sufficient number of incremental prices to determine the aggregate site demand curve. Price is increased until the estimated number of trips by all individuals or from all distance zones falls to zero. The area under the resulting second stage demand curve plus any entrance fees measures the recreation use value attributed to the resource.

The basic premise of the approach is that number of trips to use a recreation resource will decrease as the out-of-pocket and time cost of travel increase, other things remaining equal. It is assumed that all users enjoy the experience equally and that the travel cost of the most distant user represents the price intercept of all other users. Nearer users thus receive net benefits from the recreation site represented by the difference between their costs and the costs of the most distant users.

Contingent Value Approach

The contingent value approach was recently approved by the U.S. Water Resources Council as suitable for the valuation of recreation and environmental change. To apply the method, a representative sample of the affected population is asked direct questions about their maximum willingness to pay, contingent on hypothetical changes in an environmental amenity such as water quality. The value reported is assumed to correspond to the point of indifference between having that amount of income or the environmental amenity. When several values are reported contingent on changes in water quality, an inverse demand function can be estimated in which the regression coefficient for water quality provides a statistical estimate of the shift in willingness to pay with changes in water quality, all other variables in the equation remaining constant. The dependent variable is willingness to pay, and the independent variables are: number of trips; travel distance; price and availability of substitutes; characteristics of the site including water quality, congestion, and surface area; income and other socioeconomic variables.

There are a number of advantages to the contingent value approach. The method can be used to estimate the value of a specific attribute of interest. Questions can be developed to ascertain the value of the aesthetic service of a natural environment. Likewise the value of specific types of outdoor recreation such as fishing or boating can be estimated. Off-site benefits such as those to non-users can also be valued. These include preservation values associated with knowledge of the existence of a natural habitat and the satisfaction generated from bequesting an environmental amenity to future generations. Other methods of evaluation such as the travel cost approach cannot measure these values. Finally, the technique allows respondents to indicate the option value of future as well as present resource use where an imminent change may irreversibly interrupt the services provided. Thus, the method is not confined to observed behavior but can be used to estimate benefits under hypothetical prospective situations. The contingent value approach can value alternative natural resource use before changes occur. To wait until after irreversible development to value water quality would be an unnecessarily costly form of experimentation.[5]

Many economists share Freeman's[6] reservations about the approach, primarily because of the potential for strategic behavior by respondents. Such behavior may bias results in the direction of a preferred policy. A respondent may have an incentive to inflate his estimate of value of a proposed action if he feels he will not be held accountable for actual payment. He may feel that his inflated estimate will aid in ensuring the benefits of a public project while he avoids liability for any cost. Conversely, a bias may result if the respondent understates his value of the resource because he feels he cannot avoid payment for his share of the benefits of the project. He may feel confident that others will provide the necessary financial support while he enjoys the benefits, realizing that his value estimate is of little importance when aggregated with others and by itself is of little significance in affecting a policy change. In either case the respondent attempts to get a free ride from society. Utilizing market-related data has been preferred by most analysts, since such analyses are based on actual behavior rather than responses to hypothetical situations.

It is notable that objections to the contingent valuation approach have been primarily theoretical, as empirical evidence of systematic bias is at best inconclusive. Davis,[7] who pioneered the contingent value approach in a study of the recreation benefits of the Maine woods, concluded that the reported values were not significantly different from those obtained by the market-related travel cost approach. Randall and associates developed refinements in the contingent valuation technique and presented a persuasive case for its effectiveness in the valuation of environmental quality. They studied the benefits from improved air quality and other environmental amenities in the Four Corners area of New Mexico[8] and the Glen Canyon National Recreation Area.[9] They found no measurable strategic behavior by environmentalists compared to other respondents. Replication of the studies resulted in similar values. Bohm[10] conducted a controlled experiment comparing five alternative measures of willingness to pay for a public good, including actual immediate payment in cash of the stated willingness to pay. He found no significant difference in values reported by five groups each presented with an alternative willingness to pay format. Bohm[11] concluded that the theoretical objections to the contingent valuation approach could be resolved by application of an interval method. Two benefit functions would be derived, based on minimum and maximum incentives to misrepresent willingness to pay. The midpoint of the interval would represent the most acceptable value.

Unit Day Value

The unit day approach is the third method recommended by the U.S. Water Resources Council to value recreation and environmental change. The approach may be used if application of the travel cost or contingent value methods would exceed planning budget constraints and the project is small with fewer than 500,000 recreation days per year. The method relies on expert judgment to develop an approximation of the average willingness to pay for recreation use. The values selected are considered to be equivalent to consumer surplus, net of travel cost or price.

The Council guidelines classified outdoor recreation into two categories: general and specialized. General recreation activities include the majority of outdoor activities requiring the development and maintenance of convenient access and developed facilities. Included is most picnicking, tent and trailer camping, warm water boating and fishing, swimming, and small game hunting. Specialized recreation opportunities are more limited, intensity of use is low, and require more skill, knowledge, and appreciation. Specialized recreation includes trout fishing, big game hunting, upland bird and waterfowl hunting, pack trips, white water boating, canoeing, and specialized nature photography.

The Council recommended a range in value of $5.50 to $16.30 per day of specialized recreation in fiscal year 1980-81. General recreation values ranged from $1.40 to $4.10. Initially based on a survey of entrance fees at private recreation areas in 1960, unit day values have been adjusted for changes in the Consumer Price Index to the present. For example, the recommended range of specialized recreation values was $2.00-$6.00 in 1962, $3.00-$9.00 in 1973, $4.29-$12.87 in 1979, and increased to $5.50-$16.30 in 1980-81. The unit day value assigned to a particular recreation site would depend on the quality of the site.

The Council has recommended five criteria for rating the quality of recreation sites: (1) quality of the recreation experience as affected by congestion; (2) availability of substitute areas in terms of the time it would take to reach substitute sites; (3) carrying capacity in terms of adequacy of facilities for the conduct of the activity; (4) accessibility as affected by road and parking conditions; (5) environmental quality including forest, air, water, pests, climate, adjacent areas, and aesthetics of the scenery. Individual sites are rated on a 100-point scale in which the recreation experience is 30, availability of substitutes is 18, carrying capacity 14, accessibility 18, and environmental quality 20 points. Quality points for each criterion are assigned according to guidelines set forth by the Council. A value table is provided by the Council for converting the scaled values into unit day dollars representing estimated willingness to pay.

For example, specialized recreation at a site with no quality points has a value of $5.50, while a site with 50 quality points has a value of $8.20, and one with 90 quality points has a value of $14.50. A site with the maximum of 100 quality points yields a unit day value of $16.30 for specialized recreation. General recreation at a site with no quality points has a unit day value of $1.40. At a site with 50 quality points the value is $2.90, and at a site with 90 quality points the value is $3.90. A site with the maximum of 100 quality points yields a unit day value of $4.10 for general recreation. The Forest Service assigned unit day values of $8-$12 for 12-hour visitor days of wilderness recreation use in 1980, compared to $10.50 for big game hunting, $6.25 for cold water fishing, and $3 for developed and dispersed recreation use.

Recreation Benefit Studies

Walsh, Ericson, McKean, and Young[12] applied the contingent value approach to estimate three statistical benefit functions relating willingness to pay for recreation access to water quality. Tourists visiting the South Platte River Basin, Colorado, were willing to pay a $0.06 entrance fee per household recreation day to avoid each one unit decrease in perceived water quality on a scale of zero to 100, all else equal. Sample households were willing to travel 0.9 percent more to avoid each one unit decrease in water quality. They were willing to pay 1.9 percent more for waterfront property to avoid each one unit decrease in water quality. The payment vehicles were familiar methods of paying for access to water resources for recreation use. Tourist perception of water quality suitable for recreation use was based on visual attributes. Respondents ranked six color photos of varying water quality in the river basin and rated each on an index of zero to 100 with zero defined as the most polluted level and 100 clean water.

Interviews with a representative sample of 141 households visiting Rocky Mountain National Park in 1973 provided the basic data for multiple regressions estimating the relationship of willingness to pay to water quality and other socioeconomic variables. The regression equations controlled the effects of variables such as income, substitution, number of trips, leisure time, recreation activity, and other socioeconomic variables. Location and resource attributes other than water quality were controlled by assumptions independent of the regression equations. Alternative forms were tried with the linear model providing the best fit of the relationship. The proportion of variation in willingness to pay explained by the independent variables was 0.30. The usual tests of the assumptions of the model revealed no significant adverse effects on the study results. The equations and regression coefficients were significantly different from zero at the 0.01 to 0.05 level.

Oster[13] used the contingent value approach to estimate the recreation use value of water quality to residents of the Merrimack River Basin in New Hampshire and Massachusetts. The author interviewed a representative sample of 200 residents of the river basin by telephone in 1973. Average individual willingness to pay to improve water quality to naturally occurring levels was reported as $12 per year. For a typical family of three persons, this would be equivalent to $36 per year. When adjusted for inflation, these estimates are similar to recreation use benefit estimated for the South Platte River Basin.

Gramlich[14] used the contingent value approach to estimate that resident households of the Boston area would be willing to pay $21-$40 in annual taxes to improve water quality in the Charles River Basin, Massachusetts (in 1973 dollars). This is a 95 percent confidence interval around the sample average willingness to pay of $30 per household. The study was based on interviews with a representative sample of 165 households living the Boston area of the river basin during the fall of 1973. To improve water quality in other river basins throughout the United States, residents of the river basin reported willingness to pay an additional $25 in annual taxes.

Thus, total willingness to pay for improved water quality in the United States including the Charles River Basin was reported as $55 per year. The author did not ask respondents to allocate the stated willingness to pay among recreation use value and option or preservation value. The benefit estimate can be interpreted as being based on each respondent's perception of a combination of recreation use, option, and preservation demands.

The improvement in Charles River water quality was defined as from the current relatively polluted level to a level suitable for swimming but not suitable as a public water supply which was ruled out for reasons of cost. At the time of the survey, water quality in the Charles River was not sufficient for swimming, but it was clean enough for wildlife use in the upper and middle parts of the 80 mile river. Water quality tended to decline with unpleasant odor and health hazard in the heavily used lower basin area. We estimate that the perceived improvement in water quality would be equivalent to an increase from a rating of 25 to 75 on a 100-point scale with zero representing water clean enough for use as a public water supply while water presenting a health hazard would rate 100.

Ditton and Goodale[15] used a variation of the contingent value approach to estimate willingness to participate in water-based recreation activities contingent on a change in water quality in Green Bay, Lake Michigan. The data were obtained from personal interviews with a representative sample of 2,174 households living in the five county area around Green Bay, Wisconsin. Sixty-four percent of the swimmers and 54 percent of the fishermen reported they would substitute less convenient recreation sites if water quality deteriorated at Green Bay, as would 49 percent of the boaters. Thirty-one percent of the fishermen would discontinue participating as would 25 percent of the swimmers and 22 percent of the boaters. Some would continue to participate at the same location despite water pollution but would participate less, how much less was not reported. This was the case indicated by 29 percent of the boaters, 14 percent of the fishermen, and 10 percent of the swimmers. Most residents reported that the water in the bay was polluted at the time of the interview and that they had adjusted their recreation behavior as indicated.

Mathews and Brown[16] used a contingent value approach to estimate that 32 percent of salmon fishermen in the state of Washington would discontinue fishing if their favorite fishing site became polluted, and 68 percent would shift to alternative higher priced sites. The authors reported the relationship between catch of salmon and net value per day of fishing in four regions of the state of Washington: Pacific Ocean beach, Strait, Puget Sound, and fresh water rivers and lakes. Fishermen reported changes in willingness to pay contingent on number of fish caught. Fishing success is related to the quality of water and other variables. Doubling daily catch from two to four pounds of dressed fish was associated with a $13 increase in the value per day from $27 to $40. Increasing the catch by the same two pounds, from eight pounds to ten pounds per day, increased fishing value per day by only $5 from $55 to about $60. Value continues to increase for a reasonable range of catch per day up to 12 pounds, possibly higher, but at a decreasing rate.

Tihansky[17] used the unit day value approach and data on public beach use to estimate welfare losses to swimming of $27.5-$65.3 million annually from pollution along the U.S. coastline. The damages result from closed and degraded public beaches causing a reduction in swimming opportunities available. The author based his estimate on past research, concluding that a 1 percent increase in beach capacity would increase swimming activity by 0.65 percent. Improved water quality at public marine beaches was estimated to increase demand for swimming by 18.8 million activity days, or 8.5 percent. Unit day values of $1.50 were assumed for new swimmers and $3.50 for current swimmers. Whether the demand increase would be new or current swimmers accounts for the range in aggregate welfare loss estimate. No information was available on private beach swimming which is also damaged by water pollution. Private beaches account for over 90 percent of the total shoreline.

Davidson, Adams, and Seneca[18] used a U.S. participation model to predict the effect of improving water quality in the Delaware River on water-based recreation. Participation by residents of an 11-county area was projected over a 25-year planning period, 1965-1990, "with" and "without" the availability of the Delaware River below Trenton. Projected values of the independent variables for each of the 25 years were multiplied by their regression coefficients and summed. The difference between "with" and "without" projections equaled 8.7 million discounted activity days of boating and 0.9 million discounted activity days of fishing. Area of water was not significant in the regression equation for probability of participation in swimming and no estimate was made for this activity. Predicted future annual days were discounted by 5 percent annually and summed to obtain present value in days for each sport. The average number of annual days use per participant was assumed fixed at the 1960 level.

In the absence of information on travel costs, the authors did not compute the dollar benefits of water-based recreation. Rather they performed a sensitivity analysis of the effect of hypothetical benefit estimates of $1-$5 per activity day. They concluded that an average benefit of $2.55 per user day would be sufficient to equal cost of the water quality improvement program for the Delaware River (in 1965 dollars).

The University of Michigan Survey Research Center study of a representative sample of 1,352 U.S. households in 1959 provided data for the regression equations on probability of participating and number of days annually per participant household. Independent variables in the equations included: age, income, sex, education, race, children, urban location, region, coastal, water area available, and quality of facilities. The probability of participation equations explained 11 percent of the variation in the dependent variable. A number of independent variables were not significant. For example, in the probability of boating equation, the regression coefficients for income, education, and occupation were nonsignificant, while race and age squared were only marginally significant. However, since the regression equations were used to predict changes in the dependent variables, the correct procedure was to use the coefficients of all theoretically important variables, including those which were statistically insignificant, to predict changes in the dependent variables.

The critical variable, water area available, was defined as acres of recreational fishing water in each state divided by the population of the state in 1960. The regression coefficient for water area available was 0.2346 for fishing and 0.3849 for boating. This means that each acre of water per capita was associated with a 23.5 percentage point increase in the probability of participating in fishing, compared to 38.5 percent for boating. The 11-county area had 11,617 acres of water suitable for fishing without the Delaware River below Trenton, equivalent to 0.0035 acres per capita in 1960. With improved quality of 57,600 acres of the Delaware River below Trenton, the 11-county area would have 0.0208 acres per capita.

Stoevener and Stevens, et al.,[19] used the travel cost approach to estimate the change in benefit from fishing for salmon, clams, and bottomfish resulting from alternative effluent disposal plans for a Kraft paper mill near Yaquina Bay, Oregon. The authors concluded that the $22,700 annual benefit of fishing in the Bay with negligible water degradation would decline by $2,600 to $7,000 with alternative increases in effluent from the paper mill (in 1964 dollars). It was hypothesized that an increase in water pollution would damage the fishery and that this would reduce angler success which would reduce angler effort, measured by annual days of fishing.

Fishing benefits without pollution of the Bay were estimated from a multiple regression equation with annual number of fishing days per capita as the dependent variable. Independent variables were direct cost of travel and other services purchased per user day and annual household income. Changes in fishing benefits with pollution were based on two relationships. First, biologists estimated the change in fishermen success resulting from three alternative effluent disposal plans in the Bay. Second, the authors used a study by Stevens[20] which estimated the relationship between annual days of fishing and number of fish caught per day in a regression analysis of time series data from other fishing areas in Oregon. It was assumed that the elasticity of demand with respect to number of fish caught would apply to pollution caused changes in angler success in Yaquina Bay. The benefit of water quality was calculated as the area between fishing demand curves with and without pollution. Possible benefit to other recreation uses of the Bay such as boating, swimming, hiking, camping, and picnicking along the Bay were not estimated.

Reiling, Gibbs, and Stoevener[21] used a variation of the travel cost approach to estimate recreation benefits with and without improved water quality in Upper Klamath Lake, Oregon. It is the largest lake in the state with a surface area of 130 square miles, and recreation use is severely constrained by warm water temperatures and by algae growth from excessive concentrations of nitrogen and phosphorus. The authors concluded that with improved water quality, individual benefit from fishing, swimming, boating, and water skiing would increase by $10 per visit. For visits averaging 2.8 days in length, this was equivalent to benefit of about $3.50 per user day (in 1968 dollars). Number of visits was forecast to increase by 158 percent to 378,000 visits per year. Thus, recreation benefit was estimated as $3.8 million per year. In addition, the authors estimated that the increased recreation expenditure in the local economy would generate total direct and indirect sales of $2.7 million annually. However, regional impact was not included in the benefit estimate because under conditions of full employment, it represents transfer from one region to another rather than increased social benefit.

Interviews with a representative sample of 300 recreation users of four lakes, including Klamath, provided the basic data for pooled multiple regression equations in which the dependent variables were number of visits per capita and number of days per visit. Significant independent variables in the equations included: travel cost, on-site cost, income, surface water area, and an index of site characteristics. The index was based on experts' rating, using a 10-point scale, of suitability of the four sites for water-based recreation. The index was interpreted as a measure of water quality although it may have also represented variables other than water quality which would weaken the validity of the benefit estimates.

Demand without improved water quality was estimated by insertion of the average values for Klamath Lake into the pooled regression equations. To determine the number of visits with improved water quality in Klamath Lake, higher values for water quality and other variables were inserted into the regression equation and the number of visits from each distance zone was calculated as the area between days per trip demand curves with and without improved water quality. To compute benefit per visit with improved water quality, the higher values of the variables were inserted into the days per visit equation which resulted in a parallel shift in the demand curve.

Russell[22] used a travel cost demand model to estimate the effect of improved water quality in the Nashua River on participation and benefits from water-based recreation by residents of the city of Nashua, New Hampshire. The author estimated that recreation benefits from improving water quality in the particular stretch of river were $3.93 to $5.68 per capita annually, depending on assumptions about the availability of competing water area and the amount of surface area opened up by a water pollution control program. He adapted a regional travel cost demand equation for recreation at reservoirs in Texas,[23] adjusting for the shorter recreation season and increased competing water area in New Hampshire.

The Texas Parks and Wildlife Department study of a representative sample of 13,000 households in 1965 provided data on a subsample of visits to eight reservoirs by residents of counties within 100 miles. The dependent variable in the pooled multiple regression was number of visits. Independent variables included: county population, direct travel cost per trip, average county income per capita, number of alternative bodies of water within 100 miles of the county center, and water area available in each of the eight reservoirs. All variables were significant at the 0.05 level and the proportion of variation in visits explained by the variables in the equation was 0.41. The critical variable, water area available, was defined as the surface acres in a conservation storage pool. The regression coefficient for water area available in the double log function was 0.2096 which means that a 1 percent increase in water area available would result in a 0.2 percent increase in demand for water-based recreation, all else remaining unchanged.

Bouwes and Schneider[24] used the individual travel cost approach to estimate water-based recreation benefit with and without expected pollution from a storm sewer at Pike Lake in southeastern Wisconsin. It is one of many small lakes in the state less than one square mile in size which are suitable for day use, having boat launching facilities and swimming beaches with lifeguards. The authors estimated annual benefit of $20 per household, measured as the area between demand curves with and without the expected change in water quality. It was assumed that with relocation of the storm sewer, expected water quality would rise by 7 points from a rating of 10 to 3, on a 23-point scale with zero representing clean water. This was equivalent to a 30.4 percentage point change in water quality and a rate of change in benefit of nearly $0.047 per household recreation day with each one unit change in water quality on a 100-point scale. This finding is comparable to the rate of change in willingness to pay a daily fee reported by Walsh, et al., regarding tourists in the South Platte River Basin, Colorado, where the contingent value approach estimated a rate of change in benefit of $0.06 per household recreation day with each one unit change in perceived water quality on a 100-point scale.

It was possible with the information found in Bouwes and Schneider to calculate the expected shift in the travel cost demand curve with severely polluted water in Pike Lake, i.e., change from 3 to 23 on the pollution index. Demand would decline by 3.8 trips per year from 13.9 trips with a pollution index of 3 to 10.1 trips with a maximum pollution index of 23. We interpret this result to imply that recreation benefit measured as the area between demand curves with and without this level of pollution would be roughly $60 per household annually (in 1976 dollars).

On-site interviews with a representative sample of 195 households visiting eight lakes, including Pike Lake, provided the basic data for a pooled multiple regression function in which the dependent variable was number of annual visits of one day each. Independent variables in the equations included: total variable cost per trip, annual income, round-trip time, and an index of water quality. The variables included in the equation explained 20.3 percent of the variation in number of annual visits. All variables were significant at the 95-99 percent level except cost per trip, which may reflect the possibility that time is perhaps a more binding constraint than costs when considering relatively short day trips.

An important contribution of Bouwes and Schneider's study was to compare user perceptions to a technical rating of lake water quality. Recreation users at each of the eight lakes reported their perceptions of lake water quality on a 0-23 point scale, where zero represents clean water. In addition, all lakes in the state over 100 acres were rated by geologists on the basis of dissolved oxygen, disk transparency, winter kill, and algae growth. These quality variables were summed in the construction of an index with a range of 0-23 points. The effectiveness of the technical rating in predicting recreation user's perceptions of water quality was tested by regressing the average rating by recreation users of each of the eight lakes on the corresponding technical ratings. The results were encouraging with the technical ratings explaining 69.4 percent of the variation in recreation user's perceptions. The double log equation enabled the authors to predict the effect of a technical improvement brought about by a pollution control program on recreation user's perception of water quality.

Liu[25] used a variation of the travel cost demand approch to estimate the benefits to swimming from improving water quality in three lakes located in the Minneapolis-St. Paul metropolitan area from 1978 to 1979. Benefits were measured as the change in consumer surplus between demand curves with and without the perception of improved water quality. The perceived change in water quality was a dummy variable with improvement designated as 1. This variable was the basis for shifting the demand curve without improved water quality in 1978 to demand with improved water quality in 1979. It was estimated that visitors in 1979 who perceived an improvement in lake water quality which resulted from water quality improvement program expenditures in 1978, were willing to pay an additional $1.32 per person per trip to the three lakes. This was equal to total benefit of $579,000 for improved water quality in the three lakes which yielded a benefit-cost ratio of 3.3 for the lake restoration project expenditures in 1978.

On-site interviews with a representative sample of 1,261 parties visiting the three lakes, 581 in 1978 and 680 in 1979, provided the pooled data for regression equations with number of visits per year as the dependent variable. These were short day-trips with average travel costs of less than $2 per visit and an average of 10-14 trips per year. Independent variables included: travel cost, travel time cost, on-site time cost, complaints of water problems, dissatisfaction with respect to lake environment and facilities, awareness of the water quality project, trend in rate of use, perceived enjoyment of the experience, and perceived change in water quality. The linear equations explained 12-15 percent of the variation in number of trips per year to the three lakes, although a number of regression coefficients for taste and preference variables included in the equations were not significant at the 5 percent level. All strategic variables such as travel cost and perception of water quality were significant at the 5 percent level.

Sutherland[26] used a regional travel cost demand approach to estimate the benefits to boating, fishing, swimming, and camping with and without improved water quality in Idaho, Oregon, and Washington state. For example, the author estimated that improving the 7,156 miles of polluted river in the Northwest would increase recreation benefits by $26.4 million, equal to $3,683 per mile of river restored. He estimated the recreation use benefits of eight lakes as $4.40 ($2-$14) per trip for day users. This was equivalent to average benefit of approximately $35,000 per mile of lake shore suitable for recreation use. It should be noted that half of the lakes received high levels of recreation use because they were close to population centers while many rivers were in agricultural areas not conducive to recreation even if water quality were improved. Also, benefits were low, owing to an abundance of existing accessible recreation opportunities in the Northwest. In addition, the double log form of the demand function resulted in much higher use and benefit estimates than the semi-log form, the latter of which was the basis for the above estimates. Also, the estimated benefits of improved water quality were constrained by the model which held constant the total number of recreation trips to sites in the three states at 1976 levels. In reality, one would expect improved water quality to stimulate increased total water-based recreation and benefits.

Statewide recreation surveys of a representative sample of over 5,000 households in Idaho, Oregon, and Washington in 1976 provided the basic data for Sutherland's regression equations on probability of participating and number of days annually per participant household. Independent variables in the equations included: age, sex, income, accessibility, and state of residence. The probability of participating equations explained 8-25 percent of the variation in the dependent variable, compared to the days per participant equation which explained only 3-9 percent and a number of demographic variables were dropped as not significant at the 10 percent level.

The statewide surveys also provided basic data for a gravity model which was used to estimate trip interchanges between all pairs of 144 origin zones and 179 recreation areas in the region. One advantage of using the gravity model is that it permits each recreation site in the region to be a substitute for every other site as measured by travel distances (prices) and attractiveness (recreation facilities). The greater the distance from origin zones to recreation sites, the greater the probability of preferred substitutes closer to home.

A recreation site attractiveness model included activity days per year as the dependent variable and two independent variables, accessibility and facilities available. The accessibility variable was defined as a function of the number of trips from each origin zone times the likelihood that these trips will terminate at the recreation site. The facility variable was defined as the linear feet of designated swimming beach, the number of camp sites, the miles of suitable river or lake shore miles, and number of boat ramps as appropriate for each of the four activities. An acceptable recreation site was defined as one with water quality meeting the 1983 Federal water quality goals of fishable and swimmable water. How many facilities of each type which could reasonably be constructed along rivers and lakes with degraded water was obtained from state recreation officials. The estimated increment in facilities should be interpreted as the maximum potential change and not as an estimate of what would occur if water quality were improved. Recreation benefit estimates therefore represent an upper bound which could be attained only by construction of the necessary facilities.

In a preliminary EPA report,[27] Walsh[28] developed an interim procedure to estimate national recreation benefits of achieving U.S. water quality goals of 1983. Potential annual benefits to fishing, boating, nonpool swimming, and waterfowl hunting were estimated as $4.2-$10.6 billion in 1970 dollars. The substantial range in estimate reflects the uncertainty associated with limited secondary data, and should become more precise with the development of improved procedures and data.[29] Heintz, Hershaft, and Horak[30] used the same approach and data to estimate that national recreation benefits were most likely $6.3 billion in 1973 dollars. Unger[31] used substantially the same approach and data to estimate the national recreation benefits of achieving the 1977, 1983, and 1985 water quality goals with increases in population and income. More recently, Mills and Feenberg[32] used the same data and an alternative approach to estimate substantially higher national recreation benefits of achieving the water quality goals of 1985. Freeman[33] critiqued these estimates and those prepared by the National Water Quality Commission, concluding that national recreation benefits were most likely $6.7 billion with a range of $4.1-$14.1 billion in 1978 dollars.

Walsh's interim approach segmented total U.S. demand for water-based recreation into two groups. A demand curve was specified with competition or substitution and a demand curve without these factors. The aggregate demand curve with competition was equivalent to the horizontal sum of demand curves for all sites with suitable water quality for water-based recreation in 1970. Benefits would result from movement along the aggregate demand curve in response to the potential increase in supply of water suitable for water-based recreation. With substitution, benefits accrue to recreation users in the form of a decrease in expenditures on the original quantity demanded plus the consumer surplus of the increase in quantity from movement along the demand curve. The procedure recommended by the U.S. Water Resources Council to estimate the benefits of this transfer in use from existing sites to new sites of comparable quality was the travel cost saving resulting from the transfer, as demonstrated by Knetsch.[34] This avoids double counting in application of the unit day value approach during the interim period until regional demand equations can be developed with appropriate shift variables for the quality and quantity of water available.

The aggregate demand curve without competition or substitution was equivalent to the horizontal sum of demand curves for all sites without suitable water quality which nonetheless were used for water-based recreation in 1970. With improved water quality at polluted sites they continued to use, benefit was estimated as the area between their aggregate demand curves with and without improved water quality and above price which does not change.

The interim approach relied upon secondary data from a number of sources. U.S. Census studies provided information on participation in water-based recreation and average miles traveled per recreation day in 1970. A Wisconsin contingent value study estimated willingness to participate with changes in water quality. A Missouri travel cost demand study was updated to estimate average direct cost of travel per recreation day. A Colorado contingent value study estimated changes in willingness to drive and to pay with changes in water quality. The resulting national recreation benefit estimates were preliminary and tentative because they were based on studies in limited geographic locations. Responses were conditioned by existing water quality, availability of alternative recreation sites, and by the socioeconomic characteristics of the sample populations in the survey areas. Other regions of the U.S. have different water quality, substitution, and population characteristics. There is a need for regional demand studies to test the extent to which the studies in various states have provided reasonable estimates of national recreation benefits.

The National Commission on Water Quality[35] estimated the national benefits of fresh water and marine fishing, boating, and swimming at public beaches as $5.1-$6.7 billion per year (in 1978 dollars). The estimates assumed that the objectives of the Federal Water Pollution Control Act of 1972 would be attained by 1985 with projections of population, income, and other variables to that year. The national estimate was based on the following three studies funded by the Commission.

The National Planning Association[36] used a participation model to estimate that with improved water quality, fresh water fishing would increase by 26-67 million activity days per year and boating by 50-115 million activity days per year by 1985. The National Recreation Survey of 1972 provided data for the regression equations on probability of participating and number of days per participation. Independent variables in the equations included travel cost, water quality, income, race, and sex. Unit day values of $10 per fishing day and $12 per boating day were assigned by the Commission staff. The study did not estimate increased benefits to existing participants.

Battelle Memorial Institute[37] used a participation model to estimate the increase in swimming at 13 percent of the public beaches closed because of coliform bacteria contamination. Benefits resulting from projected 1985 levels of improved water quality, population, income, etc., were estimated as $191-$631 million per year. The National Recreation Survey of 1972 provided data for the regression equations on probability of participating and number of days per participant. Both equations included socioeconomic variables such as income and age, while the latter equation included a travel distance and cost variable which was interpreted as a proxy for availability of swimming beaches. Unit day values of $9 per day of increased swimming and $3 per day of swimming diverted from other beaches were assigned by the Commission staff.

Bell and Canterberry[38] used a household production function model to estimate the consumer surplus associated with marine sport fishing. They relied on secondary data from a number of sources to estimate the relationship of water quality to biological productivity and willingness to participate in fishing. On this basis, increased benefits with 1985 levels of water quality were estimated as $4 billion per year. Time as an input was valued at the foregone wage rate compared to the Water Resources Council recommended value of travel time at one-third of the wage rate. Freeman[39] estimated that this increased their benefit estimate by about one-fourth as compared to what would have resulted employing the lower value for travel time.

PRESERVATION BENEFITS FROM IMPROVED WATER QUALITY

The environmental economics literature identifies several possibilities of willingness to pay for preservation of public nonmarket aspects of environmental quality which are distinct from the direct or immediate consumer surplus benefit from use of the natural environment. These preservation benefits include option, bequest, and existence demands as outlined by Krutilla.[40] Option value is defined as the willingness to pay for the opportunity to choose from among competing alternative uses of a natural environment in the future. Existence value is the willingness to pay for the knowledge that a natural environment is preserved. Bequest value is defined as the willingness to pay for the satisfaction derived from endowing future generations with a natural environment. Each type of preservation benefit will be briefly discussed in this section. Studies which have attempted to quantify these benefits will also be reviewed.

Option Value

Weisbrod[41] originated the concept of option value. He wrote in rebuttal to Friedman's[42] advocacy of a policy of cutting down the ancient redwoods in Sequoia National Park in the event that the present value of the stream of annual net benefits accruing to park visitors was found to fall below the current commercial value of redwood lumber. Weisbrod identified two necessary conditions for the occurrence of option value: (1) infrequency and uncertainty of supply and demand for the commodity under consideration, and (2) the prohibitive high cost in time or resources of expanding production of the commodity once it has been curtailed or stopped. Visits to Sequoia National Park are usually infrequent and uncertain. Should production of the redwood forests be diverted from aesthetic enjoyment to lumbering, it would require centuries for the forests to be reestablished. According to neoclassical economic theory, the relevant opportunity cost of lumbering would be the on-site aesthetic enjoyment opportunities foregone. Weisbrod showed that considering only recreation users benefits might result in understated environmental preservation values.

Weisbrod began with a simplified analysis in which he assumed that a market existed for the collection of an admission fee from all users of Sequoia National Park. It was also assumed that the park was privately owned by a perfectly discriminating monopolist whose present value of total costs exceeded the present worth of all revenues. All external economies were assumed away. The product was considered nonstorable, and the possibility of purchase before consumption was precluded. Given these assumptions, if the private and social rates of discount are equal, then on grounds of economic efficiency the park should be closed. Its resources should be allocated to other uses.

Weisbrod, however, suggested that it may be unsound from society's standpoint to reallocate the park's resources. Given the presence of rational consumers who anticipate visiting the park, but in reality are uncertain and may or may not, such individuals may be willing to pay a fee to guarantee their future access to the park. If a private market existed whereby option value could be collected, it would influence the park owner's decision of whether or not to remain open. Without the existence of such a market, aggregate user fees will understate the total worth of the park to society. If in fact, the park closes for lack of a practical way to collect the option value, the option demands of potential future users are unfulfilled.

Weisbrod emphasized the fact that option value is significant for economic decision makers only when a decision to close the park is imminent. As long as the park remains open, the provision of the option is a costless byproduct of current operation. It fulfills the conditions of a pure collective good since all potential future users of the park can maintain the option without infringing on the consumption opportunities of others.

Following Weisbrod's introduction to option value, economists became entangled in a discussion of whether it was a totally new concept or merely "the unrecognized son of that old goat, consumer surplus."[43] Conclusive analysis by Henry[44] and Arrow and Fisher[45] showed that option value could be defined as an individual's willingness to pay for the assurance of selecting the preservation of a unique irreplaceable environmental asset facing an imminent irreversible commitment, until such time that sufficient information becomes available to choose the most beneficial alternative. In essence, what this implies is that an individual receives benefits from possessing an option to choose among various uses of a natural environment since as time passes, improved information is almost always available upon which to base a choice of the particular use that will be most beneficial. Thus, there is a greater possibility that a more appropriate decision can be made in the future because of the availability of better information at that time.

There appears to be an increasing number of decisions involving irreversible commitments of the natural environment. These decisions impose significant opportunity costs on individuals who wish to maintain an option to view endangered species, and traverse unique wilderness areas or scenic recreation sites. Unfortunately, until very recently, no empirical evidence bearing on the monetary significance of such benefits has been available to assist in the development of environmental policy. This study provides what we believe to be the first empirical estimate of option value, arising in this case from the assured choice of recreation use of preserved water quality in the presence of potentially irreversible water quality degradation due to mineral and energy development in the South Platte River Basin, Colorado.

Existence Value

Existence value has been defined as the amount an individual would be willing to pay to preserve an area as a natural habitat for the satisfaction provided by the knowledge that such an area exists. Krutilla provided a brief rationale for existence demand:

> There are many persons who obtain satisfaction from mere knowledge that part of wilderness North America remains even though they would be appalled by the prospect of being exposed to it. Subscriptions to World Wildlife Fund are of the same character. The funds are employed predominantly in an effort to save exotic species in remote areas of the world which few subscribers to the Fund ever hope

to see. An option demand may exist therefore not only among
persons currently and prospectively active in the market
for the object of the demand, but among others who place a
value on the mere existence of biological and/or geomor-
phological variety and its widespread distribution.[46]

A similar argument supporting the case of existence value is also
found in a later work by Krutilla and Fisher.[47]

Existence value is similar to option value in that it attains
relevance only when there is an imminent danger to a natural environment. Otherwise it is a pure public good, free to all as long as
the area is preserved. No one can be excluded from the satisfaction
which is derived from the area's existence.

Although an individual may not physically use a natural environment the knowledge of its existence acquired vicariously provides
utility to him. Let Q be defined as a preserved natural environment
serving as a native habitat for fish, plant, and animal life. Knowledge of the existence of this environment is defined as K. Then
$K = f(Q)$. This means that the natural environment Q is providing
the "service" of existence knowledge which yields satisfaction to
the individual. A related service provided by the area may be the
aesthetic enjoyment derived from hiking through the natural setting.
One essential difference between the two services is that the latter
is provided on-site while in the former case one need never set foot
in the area to gain satisfaction derived from the knowledge of its
existence. Since $K = f(Q)$ the satisfaction from existence knowledge
is dependent on the physical preservation of Q. The reasonable
assumption is made that the more closely an environment approximates
a natural state of existence, as opposed to a degraded environment,
the greater will be the magnitude of existence value. Willingness
to pay for existence knowledge must be added to recreation use benefits from protecting a natural environment or an underestimation of
total benefits may result.

Bequest Value

An additional type of nonuser benefit has been suggested by
Krutilla:

> We are coming to realize that consumption-saving behavior
> is motivated by a desire to leave one's heirs an estate
> as well as by the utility to be obtained from consumption.
> A bequest of maximum value would require an appropriate
> mix of opportunities to enjoy amenities experienced di-
> rectly from association with the natural environment along
> with readily producible goods. But the option to enjoy
> the grand scenic wonders for the bulk of the population
> depends upon their provision as public goods.[48]

Bequest value is the satisfaction derived from endowing future
generations with a natural environment. In many respects it is similar to existence value. Although the bequest motivation means that
nonusers may also have a desire to preserve natural environments,

there is no a priori reason that users could not likewise have such a desire with an appropriate value. Bequest value may actually be greater for users than nonusers, as the former have first-hand knowledge of the environment.

Bequest value is a pure intertemporal public good. Members of the present generation will be in a position to provide a bequest of a natural area to future generations so long as it remains preserved. If there is a threat to the area's preservation, bequest value takes on greater significance. Bequest value then is no longer a free service of the environment. It must be estimated along with use values of the environment to attain an accurate estimate of the total benefits of the preserved area. The total value of the natural environment to society may be underestimated and a serious misallocation of resources could result if bequest values are ignored.

It should be restated that as long as there is no threat to the natural environment by competing use, e.g., mineral or energy development, then option, existence, and bequest values are provided as a free public good to all who possess such satisfaction. If circumstances change so that there is a significant possibility of a competing use occurring, such values may attain great importance for society. The value of alternative uses of the environment should be ascertained to enable society to make the correct decision as to which use it should be put.

Preservation Benefit Studies

Meyer[49] surveyed residents of the Fraser River system in British Columbia. Residents were asked to assess the value of recreation and preservation related to salmon in the Fraser River. The respondents were shown the dollar expenditures per household for such services as education, social assistance, community services, etc. They were asked to provide an estimate of what they felt should be expended on these items.

Residents were then given a list of environmental and recreational services provided by the Fraser River and asked to estimate the value of these services. For residents of the upper Fraser River, these values ranged from $183 (Canadian) per household per year for the consumer surplus measure of value to $350 per household annually for a monopolist's optimal revenue measure. The latter value is determined by the revenue maximizing relationship between price and quantity of a public resource. It has been used as a measure of value when satisfactory substitutes are unavailable. Lower Fraser River resident values ranged from $436 per household for the consumer surplus measure of value to $646 per household annually for the monopolist's optimal revenue value. Finally, residents were asked to estimate what part of these values were associated with the presence of salmon in the river. Preservation values for salmon were $223 per household per year for upper Fraser River Basin residents. The annual value of recreation related to Fraser River salmon was $185.6 million. The annual preservation value was estimated as $100.6 million. Thus, total annual value of $286.2 million was increased by 54 percent with inclusion of preservation value.

A recent estimate of the nonuse benefits of water quality was prepared for the President's Council on Environmental Quality. Freeman[50] reviewed the scant evidence on nonuse benefits and concluded that national nonuse benefits of water quality were most likely $2 billion annually with a range of $1-$5 billion in 1978 dollars. The substantial range in estimate reflects the uncertainty associated with limited data, and should become more precise with the development of improved procedures and data. Nonuse benefits were defined to include preservation, option, amenity, aesthetic, ecological, and property benefits which are not directly associated with activities on or adjacent to a body of water or with diversionary uses of the water.

In an experiment using the contingent value approach, Brookshire, Eubanks, and Randall[51] queried Wyoming sportsmen on the magnitude of their recreation use, option, and existence benefits for various species of wildlife. Elk hunters were asked to estimate their maximum willingness to pay for a hunting license depending upon the number of sightings in various hunting habitats. Elk hunters were shown photos of three different hunting habitats and asked to estimate the expected number of sightings. Willingness to pay for elk hunting ranged from $30 to $152 depending on habitat and number of sightings.

Grizzly bear hunters were asked to estimate their willingness to pay an option fee for a game stamp. It was specified that revenues would be used to increase and preserve ideal habitat for bear, currently protected by a hunting moratorium. If and when the moratorium is lifted only those possessing the hunting stamp would be allowed to hunt bear. Under a second approach, big game hunters were asked to estimate the option value of increased probability of purchasing big horn sheep hunting licenses when demand exceeds supply.

Low[52] employed the contribution of time, money, and services by members of Alaska conservation organizations in support of the Alaska Lands Bill to estimate willingness to pay for the wilderness option. Members were willing to pay an average of $218 to $846 per year with the value of time in the lower estimate based on the wage value of services provided and the higher estimate based on the income of the donor. Members of special interest groups were willing to pay substantially more for preservation or option demand for wilderness at least in support of a campaign for a few years duration, than the general public would pay annually in the long run.

Walsh, Gillman, and Loomis[53] used the contingent value approach to estimate the preservation value of alternative levels of wilderness designation in Colorado. A representative sample of 218 residents of Colorado participated in a mail survey in 1980. Willingness to pay increased at a decreasing rate, contingent on the amount of wilderness area protected. Total annual preservation value to the general population increased from $14 per household for 1.2 million acres in 1980 to $19 for 2.6 million acres in 1981, to $25 for twice this amount or 5 million acres, and $32 for all potential wilderness areas estimated as 10 million acres. These total annual preservation values were allocated among option, existence, and bequest demands.

The authors used the individual travel cost demand approach to estimate average recreation use benefit of Colorado wilderness as $14 per day. This was the consumer surplus below the demand curve and above direct cost per day. The results of the study provide an empirical test and confirmation of Weisbrod's and Krutilla's proposal that the general population would be willing to pay for the preservation of wilderness resources and that these preservation or option values should be added to recreation use value to determine the total economic value of wilderness to society. In the absence of information on preservation values, insufficient public land has been allocated to wilderness protection in states such as Colorado where future resource development may irreversibly degrade environmental quality.

NOTES

1. A. Myrick Freeman III, The Benefits of Environmental Improvement: Theory and Practice, (Baltimore: Johns Hopkins University Press for Resources for the Future, 1979), 195-233.
2. Freeman, Ibid., 220-224; Frank J. Cesario, "Congestion and Valuation of Recreation Benefits," Land Economics, 56:3 (August 1980), 329-38.
3. U.S. Water Resources Council, "Procedures for Evaluation of National Economic Development (NED) Benefits and Costs in Water Resources Planning," Federal Register, 44:242 (December 14, 1979), 72, 950-65.
4. The travel cost approach was suggested by Hotelling, initially applied by Clawson and further developed by Knetsch and others. See Harold Hotelling, "The Economics of Public Recreation," The Prewitt Report, Land and Recreation Planning Division, National Park Service, (Washington, D.C.: U.S. Department of the Interior, 1949); Marion Clawson, Methods of Measuring the Demand for and Value of Outdoor Recreation, Reprint No. 10, (Washington, D.C.: Resources for the Future, February 1959); Marion Clawson and Jack L. Knetsch, Economics of Outdoor Recreation, (Baltimore: Johns Hopkins University Press, 1966).
5. David S. Brookshire and Thomas Crocker, "The Use of Survey Instruments in Determining the Economic Value of Environmental Goods: An Assessment," in Assessing Amenity Resource Values, Rocky Mountain Forest and Range Experiment Station Report No. 68, (Fort Collins: U.S. Forest Service, 1979).
6. Freeman, op cit., 85-105.
7. Jack L. Knetsch and Robert K. Davis, "Comparison of Methods for Recreation Evaluation," in Water Research, Allen V. Kneese and Stephen C. Smith, eds., (Baltimore: Johns Hopkins University Press, 1966).
8. Alan Randall, Barry Ives, and Clyde Eastman, "Bidding Games for Valuation of Aesthetic Environmental Improvement," Journal of Environmental Economics and Management, 1 (Fall 1974), 132-49.
9. David S. Brookshire, Barry C. Ives, and William D. Schulze, "The Valuation of Aesthetic Preferences," Journal of Environmental Economics and Management, 3 (Fall 1976), 325-46.

10. Peter Bohm, "Estimating the Demand for Public Goods: An Experiment," European Economic Review, 3 (June 1972), 111-30.

11. Peter Bohm, "Estimating Willingness to Pay: Why and How?" Scandinavian Journal of Economics, 81 (1979), 142-53.

12. Richard G. Walsh, Ray K. Ericson, John R. McKean, and Robert A. Young, Recreation Benefits of Water Quality: Rocky Mountain National Park, South Platte River Basin, Colorado, Technical Report No. 12, Colorado Water Resources Research Institute, (Fort Collins: Colorado State University, May 1978).

13. Sharon Oster, "Survey Results on the Benefits of Water Pollution Abatement in the Merrimack River Basin," Water Resources Research, 13 (December 1977), 882-84.

14. Fred W. Gramlich, "The Demand for Clean Water: The Case of the Charles River," National Tax Journal, 30:2 (June 1977), 183-94.

15. Robert Ditton and Thomas Goodale, Marine Recreational Use of Green Bay: A Survey of Human Behavior and Attitude Patterns, Technical Report No. 17, Sea Grant Program (Madison: University of Wisconsin, 1972).

16. B. S. Mathews and Gardner S. Brown, Economic Evaluation of the 1967 Salmon Fisheries of Washington, Technical Report No. 2, Washington Department of Fisheries (Olympia, Washington, 1970).

17. Dennis P. Tihansky, "Recreational Welfare Losses from Water Pollution Along U.S. Coasts," Journal of Environmental Quality, 3:4 (October-December 1974), 335-46.

18. Paul Davidson, F. Gerard Adams, and Joseph Seneca, "The Social Value of Water Recreational Facilities Resulting from an Improvement in Water Quality: The Delaware Estuary," in Water Research Allen V. Kneese and Stephen C. Smith, eds., (Baltimore: Johns Hopkins University Press, 1966), 175-211.

19. Herbert H. Stoevener, J. B. Stevens, H. F. Horton, Adam Sokoloski, L. P. Parrish, and E. N. Castle, Multi-Disciplinary Study of Water Quality Relationships: A Case Study of Yaquina Bay, Oregon, Special Report 348, (Corvallis: Oregon State University, February 1972).

20. Joe B. Stevens, "Recreation Benefits from Water Pollution Control," Water Resources Research, II (Second Quarter 1966), 167-82.

21. S. D. Reiling, K. C. Gibbs, and H. H. Stoevener, Economic Benefits from an Improvement in Water Quality, (Washington, D.C.: Environmental Protection Agency, January 1973).

22. Clifford S. Russell, "Municipal Evaluation of Regional Water Quality Management Proposals," in Models for Managing Regional Water Quality, Robert Dorfman, Henry D. Jacoby, and Harold A. Thomas, Jr., eds., (Cambridge, Mass.: University Press, 1972), 142-203.

23. Herbert W. Grubb and James T. Goodwin, Economic Evaluation of Water-Oriented Recreation, Report No. 84, (Austin: Texas Water Development Board, September 1968).

24. Nicolaas W. Bouwes and Robert Schneider, "Procedures in Estimating Benefits of Water Quality Change," American Journal of Agricultural Economics, 61:3 (August 1979), 535-39.

25. Ben-chieh Liu, "Recreational Benefit Estimation for Lake Water Quality Improvement: A Comparative Analysis," Annual Conference of the American Water Resource Association, (Minneapolis,

October 1980).

26. Ronald J. Sutherland, A Regional Recreation Demand and Benefits Model, Draft Report, Environmental Research Laboratory, (Corvallis, Ore.: U.S. Environmental Protection Agency, September 1981).

27. Fred H. Abel, Dennis P. Tihansky, and Richard G. Walsh, National Benefits of Water Pollution Control, Preliminary Draft, (Washington, D.C.: Environmental Protection Agency, 1975).

28. Richard G. Walsh, "Recreational User Benefits from Water Quality Improvement," Outdoor Recreation: Advances in Application of Economics, General Technical Report WO-2, Forest Service, (Washington, D.C.: U.S. Department of Agriculture, 1977), 121-32.

29. A. Myrick Freeman III, "Benefits of Pollution Control," in Critical Review of Estimating Benefits of Air and Water Pollution Control, A. Hershaft, ed., Report to the U.S. Environmental Protection Agency, (Rockville: Enviro Control, Inc., October 1976), 11.1-11.55.

30. H. T. Heintz, A. Hershaft, and G. C. Horak, National Damages of Air and Water Pollution, Report to the U.S. Environmental Protection Agency, (Rockville: Enviro Control, Inc., 1976).

31. Samuel G. Unger, National Benefits of Achieving the 1977, 1982, and 1985 Water Quality Goals, (Manhattan, Kansas: Development Planning and Research Associates, Inc., April 1976).

32. Edwin Mills and Daniel Feenberg, Measuring the Benefits of Water Pollution Abatement, Report to the U.S. Environmental Protection Agency, (Washington, D.C., 1979).

33. A. Myrick Freeman III, The Benefits of Air and Water Pollution Control, A Review and Synthesis of Recent Estimates, Report prepared for the Council on Environmental Quality, (Washington, D.C., 1979).

34. Jack L. Knetsch, "Displaced Facilities and Benefit Calculations," Land Economics, 53:1 (February 1977), 123-29.

35. National Commission on Water Quality, Staff Report, (Washington, D.C., 1976).

36. National Planning Association, Water-Related Recreation Benefits Resulting from P.L. 92-500, (Washington, D.C., 1975).

37. Battelle Memorial Institute, Assessment of the Economic and Social Implications of Water Quality Improvements on Public Swimming, (Columbus, Ohio, 1975).

38. Frederick W. Bell, and E. Ray Canterberry, An Assessment of the Economic Benefits Which Will Accrue to Commercial and Recreational Fisheries from Incremental Improvements in the Quality of Coastal Waters, (Tallahassee: Florida State University, 1975).

39. Freeman (1979), op. cit., 204-9.

40. John V. Krutilla, "Conservation Reconsidered," American Economic Review, 57 (September 1967), 777-86.

41. Burton A. Weisbrod, "Collective-Consumption Services of Individualized-Consumption Goods," Quarterly Journal of Economics, 78 (August 1964), 471-77.

42. Milton Friedman, Capitalism and Freedom, (Chicago: University of Chicago Press, 1962).

43. Millard F. Long, "Collective-Consumption Services of Individual Consumption Goods: Comment," Quarterly Journal of Economics,

81 (May 1967), 351.

44. Claude Henry, "Option Values in the Economics of Irreplaceable Assets," The Review of Economic Studies: Symposium on the Economics of Exhaustible Resources, (1974), 89-104.

45. In an article published at about the same time, Arrow and Fisher demonstrated the possibility of quasi-option value of individuals and society. Kenneth J. Arrow and Anthony C. Fisher, "Environmental Preservation, Uncertainty, and Irreversibility," Quarterly Journal of Economics, 88 (1974), 312-19.

46. Krutilla, op. cit., 781.

47. John V. Krutilla and Anthony C. Fisher, The Economics of Natural Environments, (Baltimore: Johns Hopkins University Press for Resources for the Future, 1975), 11-15, 22.

48. Krutilla, op. cit., 784.

49. Phillip A. Meyer, Recreational and Preservation Values Associated with the Salmon of the Fraser River, Information Report Series No. PAC/IN-74-1, (Vancouver, B.C.: Environment Canada, Fisheries and Marine Service, Southern Operations Branch, Pacific Region, 1974).

50. Freeman (1979), op. cit., 135, 171.

51. David S. Brookshire, Larry S. Eubanks, and Alan Randall, "Valuing Wildlife Resources: An Experiment," Transactions, North American Wildlife Conference, 38 (1978), 302-10.

52. Christopher R. Low, "The Option Value for Alaskan Wilderness," Ph.D. dissertation, (Los Angeles: University of California, 1979).

53. Richard G. Walsh, Richard A. Gillman, and John B. Loomis, Wilderness Resource Economics: Recreation Use and Preservation Values, Department of Economics, (Fort Collins: Colorado State University, 1981).

3
The South Platte River Basin, Colorado

The South Platte River Basin, Colorado, was selected as the study area. The river basin is part of the Missouri River drainage system. It drains an area of 19,450 square miles in northeastern Colorado, approximately one-fifth of the total land area in the state. Figure 3.1 shows that it extends from the Continental Divide on the west to the Nebraska border on the east, and from the Wyoming border on the north to a boundary extending southwesterly to Colorado Springs. The river basin includes an area of relatively dense population known as the Northern Front Range of Colorado.

SELECTION CRITERIA

Selection of the basin as a study area was based on four criteria:

1. Major Population Center. With a population of approximately 1,717,900 in 1976, the river basin accounted for two-thirds of Colorado's estimated population of 2,628,127. The Denver metropolitan area is located in the river basin and with a population of about 1.4 million is similar to other large cities in the United States. Water quality improvement benefits to residents of Denver should generally reflect benefits to residents of other major urban areas in the U.S.
2. Major Water Recreation Area. There is a substantial demand for water-based recreation within the river basin. Water contact recreation activities are particularly large. An estimated 7 million water-based recreation activity days or 40.5 percent of the state total of 17.5 million days were spent in the South Platte River Basin in 1971. Primary activities include swimming, waterskiing, and fishing, all of which are dependent on a high level of water quality. As the population continues to expand, water-based recreation within the basin will also increase.
3. Water Quality. Water quality is generally poor in the river basin. The South Platte River Basin contains most types of effluent associated with modern economic development. Urban

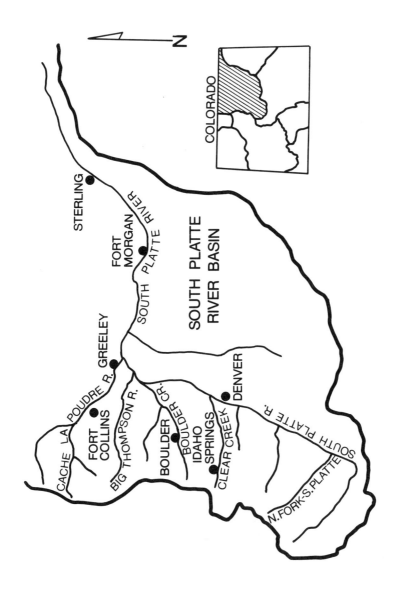

Figure 3.1. The South Platte River Basin, Colorado.

runoff, irrigation return flow, mine drainage, treated and untreated sewage, feedlot runoff, industrial wastes, and natural pollutants are common effluents in the river basin. The level of water quality degradation varies with the source of pollution, the waste load a stream accepts, and population density.
4. <u>Imminent Mining Expansion</u>. Mining activity is expanding with the opening of the Henderson molybdenum mine west of Denver, newly found ore deposits, the opening of abandoned metal mines, and increased interest in open-pit coal mining. This expansion is likely to have a pronounced effect on the river basin's water quality. There are more than 250 identifiable minerals in the South Platte River Basin. Metallic minerals include gold, silver, copper, lead, zinc, tungsten, molybdenum, and uranium. Two coal fields are also located in the basin. The Denver Region coal field is an area of 535 square miles extending from the Colorado-Wyoming state line southward to the Arkansas River Basin. The South Park coal field covers about 100 square miles in Park County. Over 3.7 billion tons of coal is estimated to exist in these two fields.

Many types of water-based recreation such as fishing and swimming are not compatible with water pollutants from mine drainage and other sources. Since current recreation demand is large relative to the available supply of water-based recreation areas, the consequences of further degradation may be of a much larger order of magnitude than if demand were small and available supply large.

ENVIRONMENTAL AND ECONOMIC CHARACTERISTICS

The South Platte River Basin lies in two major physiographic regions, the Rocky Mountains and the Great Plains. Streams originate in the higher mountainous elevations with pristine water quality, and gradually yield to degradation as elevation diminishes and human encroachment increases. There are some areas within the basin even at high elevations where streams have become heavily polluted through past mineral mining, milling operations, and abandoned shaft drainage. The Central City and Boulder Creek drainage areas are examples.

Rocky Mountain National Park on the northwestern edge of the river basin contains the headwaters of the Cache la Poudre River, the Big Thompson River, and the St. Vrain River, major northern tributaries to the South Platte River. The river basin contains 2,400 miles of fishing streams, about 30 percent of the 8,233 miles of streams in the state capable of sustaining game fish such as trout.

Denver is the commercial center of the Rocky Mountain region and has recently experienced a rapid increase in population as a result of the shift in population from the northeastern to the southwestern United States along with an acceleration in energy exploration and development. Smaller cities in the river basin include Fort Collins, Boulder, Greeley, Longmont, Loveland, and Sterling. These cities contribute to the urban setting of the river basin.

Major sectors of employment in the river basin include manufacturing, service industries, education, and government. The 1969 Census of Agriculture showed approximately 10,158 farms in the South Platte River Basin. With an average family size of four persons, farm population was estimated at 40,632 or 2.4 percent of total population in the river basin. About 11.5 percent of the residents of the river basin live in areas of less than 2,500 people. A substantial portion of the farm crops produced in the river basin are irrigated, since most of the river basin east of the mountains is arid and irrigation is necessary for agricultural production other than wheat and grazing. Heavier rainfall and snow occurs at the higher elevations. Runoff in the spring is captured in reservoirs for agricultural and domestic consumption. Water is also transferred into the river basin from the Colorado River Basin to the west.

The South Platte River Basin contains 267 lakes and reservoirs suitable for shoreline fishing, about 38 percent of the 711 lakes and reservoirs in the state. With 1,122 miles of shoreline, the river basin contains 48 percent of the 2,314 total miles of lake and reservoir shoreline in the state. The river basin provides 40.5 percent of water-based recreation activities in the state. It further accounts for 58 percent of the lake swimming, 56 percent of the sailing, and 49 percent of the power boating. It provides 34 percent of the lake fishing and 26 percent of the stream fishing. Resident water-based recreation use of the river basin accounted for approximately 72 percent of the total outdoor recreation use by residents and nonresident tourists.

WATER QUALITY

The Colorado Water Quality Control Act of 1973 required water in the state to be maintained and enhanced for the propagation of wildlife and aquatic life and for domestic, agricultural, industrial, recreational, and other beneficial uses. Colorado Water Quality Standards were established in 1974. The Standards[1] require that all water in Colorado: (1) will be free from discharges which form sludge, floating materials, materials that produce color or odor or create a nuisance, toxic substances, oil, or grease; (2) radioactive surface waters will be maintained at the lowest practical level and will not exceed Federal Drinking Water Standards; and (3) salinity concentrations will be maintained or reduced.

State waters are classified into Class A and Class B streams. Class A waters will be suitable for all beneficial uses including primary water contact recreation. Class B waters will be suitable for all purposes except primary water contact recreation.

Degradation of water quality in the state is caused by two major sources of effluent: (1) point pollution such as treated sewage from wastewater treatment facilities and industrial dischargers such as sugar beet processing plants and, (2) nonpoint pollution such as urban runoff and oxidized surface minerals which flow as runoff into mountain streams.

Almost all tributary streams to the South Platte River are classified as Class B. Lakes and reservoirs are classified as Class A. Streams in the river basin are generally out of compliance with state water quality standards. Major point sources of pollution are municipal and industrial discharges. In 1974, the Colorado Department of Health and local health departments found 95 of approximately 120 discharges out of compliance with state effluent standards.[2] The largest nonpoint source of pollution in the basin is irrigation return flow. Feedlot runoff ranks second as a source of nonpoint pollution. Although mine drainage is generally limited to streams in the headwaters of the basin there is a total of 111 miles of streams polluted with heavy metals. It is estimated that a 90 percent reduction in heavy metals pollution would be required before fish could live in Clear Creek above Denver. Urban runoff, septic tanks, leachate from dumping areas, and construction activities are other significant sources of nonpoint pollutants.

Although a number of parameters might have been selected to develop an index for water quality, we decided to use heavy metal content as the water quality index in this study. There are several important reasons for this decision. They include: (1) over 111 miles or 5 percent of the streams in the river basin were polluted with heavy metals[3] generating economically significant social costs, (2) this particular form of pollution tends to be irreversible[4] because of the prohibitive costs of water quality improvement in many areas, (3) a quantitative measure of water quality facilitates economic analysis of the benefits and costs of any proposal to improve water quality, (4) the visual effect of heavy metal effluent can be shown by color photos which is an essential part of the contingent value approach used in this study, and (5) recent data was available on the heavy metal content of water in the river basin.

A complete biological analysis of water quality in the South Platte River Basin was unavailable at the time of this study. Furthermore, no overall index of water quality was employed in Colorado at that time. The impact of heavy metal effluent on water-based recreation and aesthetics is closely associated with many other types of water pollution. Therefore, heavy metal content serves as a representative water pollutant.

CONSEQUENCES OF MINING IN COLORADO

The mountainous area referred to as the Colorado Mineral Belt contains most of the state's ore deposits and mining activity. This region extends from Boulder on the eastern slope of the Rockies to the south and west of Durango in the San Juan Mountains. Metallic ores typically found in this region include copper, zinc, lead (called base metals), gold, and silver (precious metals). Also common to the region are deposits of arsenic, cadmium, chromium, cobalt, iron, manganese, mercury, molybdenum, nickel, selenium, vanadium, and some others. Most often these metals are found not in free form but as an ore.

Mining has had a significant impact on Colorado's economy. The total value of metallic mineral production in Colorado was $245.5 million in 1975. Table 3.1 shows a substantial increase in the nominal dollar value of Colorado metallic mineral production from 1973 to 1975. The state had 223 active mines employing 6,437 people in 1975.[5] Much of this mining activity occurs within the South Platte

Table 3.1. Production of Metals in Colorado, 1973-75.

Metallic Minerals	Colorado Metals Production (Nominal Dollars Per Year)		
	1973	1974	1975
Molybdenum	$ 96,654,249	$124,015,562	$146,636,557
Zinc	15,890,102	25,405,074	28,313,631
Uranium	7,508,996	12,228,804	12,585,840
Silver	8,764,824	11,561,032	12,512,045
Tungsten	6,931,270	9,129,943	10,243,190
Gold	6,177,731	7,685,361	10,112,265
Vanadium	4,874,688	11,600,362	9,835,382
Lead	7,596,107	9,416,993	9,152,549
Copper	3,312,705	4,876,326	4,198,791
Cadium	396,186	634,247	618,048
Iron	1,058,574	1,012,538	692,087
Tin	490,131	647,770	611,849
Miscellaneous Metallics		53,833	9,800
Total Metallic Mineral Production	$159,655,563	$218,267,845	$245,522,034

Source: Colorado Division of Mines, A Summary of Mineral Industry Activities in Colorado 1975 (Denver: State of Colorado, Department of Natural Resources, 1975), p. 14.

River Basin.

Metallic ore production will most likely continue to be an important sector of the state's economy in the future. The production of molybdenum which leads all other metals is continuing to expand with opening of the large Henderson molybdenum mine and mill near Berthoud Falls in the South Platte River drainage system. Furthermore, with the deregulation of the price of gold it has become economically feasible to reopen some of the inactive gold and silver mines in the state. There is also some evidence that new deposits of base and precious metals might be found in the San Juan Mountains.[6]

Although metal mining has been beneficial to the state's economy, development of the industry has not proceeded without social costs. One consequence of a flourishing minerals industry is the inherent problem of tailings and mine shaft drainage into the waterways of the state. Since metals found in ore form are usually combined with other minerals, the sought-after metal must be extracted through milling. According to Wentz:

> Theoretically, the degradation of water quality so often associated with metal- and coal-mining operations would not manifest itself if the metal-sulfide minerals were allowed to remain in the reducing environments under which they were formed. Problems arise only when these minerals become oxidized as when they are transported

to the earth's surface.[7]

Morgan and Wentz explained:

> Draining mines and tailings piles are sources of concern because of the acid and metals they release. The acid results from the breakdown of pyrite (FeS_2) when it is exposed to oxidizing conditions. In this process, ferrous (Fe^{+2}) ions are released and oxidized to the ferric state (Fe^{+3}). The ferric ions hydrolyze, thus forming relatively insoluble ferric hydroxide ($Fe(OH)_3$), which precipitates on the stream bottom. Other metal ions may be absorbed on or coprecipitated with the $Fe(OH)_3$, thereby forming a metal-rich coating (yellow boy) on the rocks in affected streams. The overall process can be summarized as follows:
>
> $$FES_{2(s)} + \frac{15}{4} O_2 + \frac{7}{2} H_2O \rightarrow Fe(OH)_{3(s)} + 2SO_4^{-2} + 4H^{+1}.$$
>
> Metal sulfides other than pyrite (for example, sphalerite (ZnS) and galena (PbS)) will be broken down by the acid released in the above reaction. Their dissolution releases metals to the water but does not result in the production of additional net acid.[8]

It is important to point out that this oxidation process may also occur naturally as traces of iron and other metals may often be found near the earth's surface and washed into stream beds. Many streams in the Colorado Mineral Belt would contain trace elements of these metals whether or not mineral mining occurs. To what degree mining or naturally caused oxidation is responsible for water quality degradation is not known. What is clear is that mining has contributed significantly to the problem of water quality.

Coal mining in Colorado may not cause a similar problem since it has a low sulfur content.[9] This reflects the low pyritic sulfur content, an essential compound in the formation of mine acid. The full consequences of open-pit coal mining in the western United States are uncertain. Research results at this time are inconclusive. There does appear to be some evidence that elevated salt loading has occurred over extended periods after reclamation of mined areas. This is of special concern since total dissolved solids are generally already high. Elevated salt levels may make water unsuitable for human and animal consumption. Water high in salt content also cannot be used for irrigation.

Chief among the areas of environmental concern associated with mine drainage are: (1) the toxic effects of acids and metals on aquatic life, the adjacent animal population, and on human beings, and (2) losses of aesthetic and recreation values. Related problems include: corrosion of structures in the contaminated water along with industrial limitations of use, and restricted use for food processing and irrigation.[10] Additionally, there is an associated loss of bacteria which is required for biodegradation and self-purification. Hence there is an increased impurity load caused by the toxic metals. Concern will be focused here on the first two of these simultaneous problems.

The effects of toxicity may be best viewed on a spectrum from chronic ailments to immediate death. Most trace elements are beneficial and often necessary for life, yet taken in large quantities all may be toxic. Human beings can usually escape the hazardous effects of the metallic elements through avoiding contact and by filtration of drinking water. Aquatic plant life will assimilate trace elements in streams which are in turn used by higher aquatic organisms for food. Concentrations of these trace elements may eventually become large enough to be toxic to the higher aquatic organisms. The adjacent animal community may have a concentration of these elements in their systems from ingesting contaminated water or aquatic organisms taken as food. Antagonistic and synergistic effects cause complications in determining specific tolerances of the various trace elements found in the streams. Different chemical structures of the same element can further compound the toxic effects on various life forms.

Standards for safe tolerance of trace elements currently used by the Public Health Service, the State of Colorado, and the World Health Organization of the United Nations are listed in Table 3.2. Since Colorado has not established tolerances for fish and wildlife, Wentz has summarized recommended standards for Colorado Class B waters which are shown as Column 3 of Table 3.2. It should be noted that these standards are of a tentative nature and subject to revision.

The loss to recreation and aesthetics results from discoloration and toxicity to aquatic organisms and wildlife. Polluted water becomes devoid of fishlife so that fishing benefits are negligible. Swimming benefits are lost since skin contact results in irritation and ingestion of unfiltered water is toxic. Benefits from boating, aesthetic appreciation, camping, and as a source of drinking water are minimal or lost altogether.

Irreversible water quality conditions exist in several areas of the upper South Platte River Basin as a result of past and current mining practices and the prohibitive cost of rectifying the damage. Irreversible drainage flows from abandoned and active mine shafts, mill sites, slag piles, and tailing ponds.

Table 3.2. Comparative Drinking Water and Biological Standards for Trace Elements.

Water Quality Parameter	Government Agency		Recommended Tolerances for Colorado Fish and Wildlife
	U.S. Public Health Service and the State of Colorado	World Health Organization	
	(micrograms per liter)		
Aluminum	-------	-------	-------
Antimony	-------	-------	-------
Arsenic	50,^{a} (10)^b	200^a	1,000
Barium	1,000^a	-------	-------
Bismuth (III)	-------	-------	-------
Bismuth (IV)	-------	-------	-------
Cadmium	10^a	50^a	10
Chromium	50^a	50^a	-------
Cobalt	-------	-------	-------
Copper	1,000^b	3,000^b	10-20
Cyanide	200,^{a} (10)^b	10^a	-------
Fluoride	(c)^b	1,500^b	-------
Iron	300^b	100^b	300
Lead	50^a	100^a	5-10
Manganese	50^b	100^b	1,000
Mercury	5^a	-------	-------
Molybdenum	-------	-------	-------
Nickel	-------	-------	50
Selenium	10^a	50^a	1,000
Silver	50^a	-------	-------
Titanium	-------	-------	-------
Uranium	-------	-------	-------
Vanadium	-------	-------	-------
Zinc	5,000^b	5,000^b	30-70

Source: Dennis A. Wentz, Effect of Mine Drainage on the Quality of Streams in Colorado, 1971-1972, Colorado Water Resources Circular No. 21, (Denver: Colorado Water Conservation Board, 1974), p. 26.

^{a}Maximum permissible concentration.
^{b}Recommended limit.
^{c}Varies inversely with average annual maximum daily air temperature.

NOTES

1. Water Quality Control Division, *Colorado Water Quality Report, 1975*, (Denver: Colorado Department of Health, 1975), 8-11.
2. Ibid., 22.
3. Dennis A. Wentz, *Effect of Mine Drainage on the Quality of Streams in Colorado, 1971-1972*, Colorado Water Resources Circular No. 21, (Denver: Colorado Water Conservation Board, 1974), 1.
4. This is a necessary prerequisite for the existence of option value. The authors are aware that mining technology has improved markedly in the recent years and state discharge laws are much stricter so that this may not be true in all cases.
5. Colorado Division of Mines, *A Summary of Mineral Industry Activities in Colorado, 1975*, (Denver: Colorado Department of Natural Resources, 1975), 14.
6. Wentz, op cit., 12.
7. Ibid., 15.
8. Robert E. Morgan and Dennis A. Wentz, *Effects of Metal-Mine Drainage on Water Quality in Selected Areas of Colorado, 1972-1973*, Colorado Water Resources Circular No. 25, (Denver: Colorado Water Conservation Board, 1974), 6.
9. Wentz, op cit., 102.
10. Ibid., 25.

4
Data Collection Procedures

This chapter describes the data collection procedures used in the study. The chapter opens with a discussion of the procedure employed in selection of the sample. A comparison of population and sample demographic characteristics is also included to determine possible sources of sample bias. Finally, a brief discussion of survey interviewing procedures is presented.

SAMPLE SELECTION

A random sample of residents was selected from the Denver metropolitan area and from Fort Collins, a smaller university town 65 miles north of Denver. The Denver area has the largest population base in the South Platte River Basin and is representative of other large cities in the U.S. Fort Collins was selected as representative of smaller cities within the basin such as Boulder, Greeley, Loveland, and Sterling.

A total of 202 in-depth interviews were completed in the two cities from May through July 1976. The 101 interviews in the Denver metropolitan area, excluding Boulder County, represented a very small proportion of the total 424,900 households, approximately 1 in 4,200. Boulder County was excluded from the Denver survey area on the supposition that it is more representative of the smaller cities within the basin. The 101 interviews in Fort Collins included about 0.5 percent of the 18,923 households in the city. There were approximately 58,531 residents of Fort Collins in 1976. The number of households interviewed was based on experience with similar surveys and was consistent with recommendations of the U.S. Water Resources Council that a minimum of 200 households be sampled in contingent value studies of recreation and environmental quality.

The sample was randomly selected from up-to-date telephone directories for both cities. Denver and Fort Collins are highly mobile communities and it was therefore necessary to have a current residential listing. It is recognized that sampling from telephone directories precluded the selection of residents who moved into the cities within the year and also those without phone service or published listings. Bias from these sources was considered small since 92 percent of households in the Denver metropolitan area and

95 percent of households in Fort Collins had phone service.[1] Telephone directories were more current than city directories which were over a year old.

For the Denver metropolitan area, one name was drawn randomly from every ninth page of the telephone directory. Two names were drawn randomly from each page of the telephone directory for Fort Collins. Each household selected was sent an introductory letter describing the study. The interviewers contacted respondents by telephone to arrange appointments for an interview, proceeding through the initial list until the pre-established quota of 100 interviews was completed for each city.

Table 4.1 shows the rate of acceptance and refusal in the two cities. In Denver, 25.8 percent of an initial list of 392 households were interviewed, compared to 48.6 percent of 208 households in Fort Collins. The refusal rate was 21.7 percent in Denver and 16.3 percent in Fort Collins. A relatively large proportion of the residents could not be contacted although telephoned at least twice. Interviewers were unable to contact 42.1 percent of the potential respondents in Denver and 31.7 percent in Fort Collins. Letters returned by the Postal Service as nondeliverable accounted for 10.4 percent of the sample in Denver and 3.4 percent in Fort Collins.

Table 4.1. Sample Response in Denver and Fort Collins, Colorado, 1976.

Response	Denver Metropolitan Area		Fort Collins		South Platte River Basin	
	Number	Percent	Number	Percent	Number	Percent
Total Sample	392	100.0	208	100.0	600	100.0
Accepted and Interviewed	101	25.8	101	48.6	202	33.7
Refused	85	21.7	34	16.3	119	19.8
Could Not Contact	165	42.1	66	31.7	231	38.5
Returned Letters	41	10.4	7	3.4	48	8.0

CHARACTERISTICS OF SAMPLE AND POPULATION

To test for possible bias, demographic characteristics of the sample were compared with the reported characteristics for the resident population. For most comparisons, the sample statistic was very close to the population parameter. Table 4.2 lists the demographic data for the population and sample.

The metropolitan Denver sample under-represented young residents of 18-24 years of age. There is less chance that younger residents will be listed individually in telephone directories. Often younger adults continue to live with their families where the listing is not in their name. If they are living independently of their families they are likely to be living in a shared household so that the listing may be under another individual's name.

Table 4.2. Comparison of Population and Sample Demographic Profiles for Denver and Fort Collins, Colorado, 1976.

Statistic	Denver		Fort Collins	
	Population	Sample	Population	Sample
	Percent (Unless Otherwise Indicated)			
Male/Female Ratio[a] (1975)	49.5/50.5	48.9/51.1	53.1/46.9	63.3/36.4
Median Age[a] (1970) (18 years and over)	38.2	46.5	40.5	38.0
Age Distribution[a] (1970)				
18-24	19.7	7.6	23.2	9.1
25-49	48.9	50.0	48.5	57.6
50-64	19.5	25.3	16.8	16.1
65 and over	11.9	17.4	11.6	17.2
Median Education[a] (1970) (25 years and over)	12.5	14.6	12.6	14.6
Race[b] (1976)				
White	79.7	94.6	91.2	99.0
Other	20.3	5.4	8.8	1.0
Median Income	$14,647[a]	$14,958	$13,500[c]	$12,838
Income Distribution[a] (1975)				
(Under $5,000)(Under $6,000)	8.2	14.1	11.1	15.2
($5,000-$7,999)($6,000-$8,499)	8.0	8.7	13.5	10.1
($8,000-$9,999)($8,500-$10,999)	9.1	9.8	11.0	12.1
($10,000-$14,999)($11,000-$15,999)	26.5	22.8	25.9	31.3
($15,000 & over)($16,000 & over)	48.2	44.6	38.5	31.3

Notes to Table 4.2. are on p. 48.

Notes to Table 4.2.

[a]Taken or calculated from: Colorado Department of Health, Records and Statistics Section, Statistics Unit, Demographic Profile: Colorado Planning and Management Districts 2 and 3, (Denver: Colorado Department of Health, May 7, 1976), pp. 1-3, 7.

[b]Colorado Division of Planning, Demographic Section, Ethnic Group Population of Colorado Counties, 1960-1976, (Denver: Colorado Division of Planning, April 23, 1976), pp. 1, 2, 12.

[c]Estimate for the Fort Collins Metropolitan Statistical Area by the Department of Housing and Urban Development, Denver, December 31, 1976.

Minorities including Spanish surnamed, black Americans, and American Indians were also under-represented. A large percentage of minority residents with common Spanish surnames are concentrated on a relatively few pages of the alphabetically listed directories. The selection procedures used in this study would result in an under-representation of minorities because of this phenomenon. It may also be true that a smaller proportion of minority groups have telephone service. Twenty percent of the population of the Denver metropolitan area were nonwhite as compared to 5 percent of the sample.

Median family incomes were virtually identical for the sample and the population at about $14,000. Income distribution categories for the sample had slightly different brackets than the population data brackets. However, it appears that the sample included slightly more lower income people than in the general population.

Median education of the sample at 14.6 years was higher than for the population which was 12.5 years in 1970. This may have resulted for three reasons. Education levels may have increased since 1970. Respondents also may have included years of vocational and technical training as part of reported formal education. It may also be possible that more highly educated individuals consented to be interviewed.

Sample representation of the sexes closely approximated the population characteristics. There was some under-representation of women since husbands often served as the spokesman for the family. Median age for the sample was 38 years compared to a 40.5 year median age for the population.

ESTABLISHING CONTACT WITH RESPONDENTS

The method used to contact potential respondents was adopted from a study by Meyer.[2] An introductory letter was mailed to respondents about seven days before they were contacted by phone. The introductory letter stated that residents would be contacted and asked to participate in a survey of attitudes pertaining to the quality of water resources in Colorado. It stated that there was no obligation to cooperate in the survey but that those who did might influence future water quality decisions. A copy of the introductory letter is shown in Appendix B. The letter of introduction proved useful in identifying the interviewers as employees of Colorado State University which added credibility to the survey. This was important as the public is subjected to many telephone sales approaches.

Approximately one week after mailing the introductory letter, interviewers began telephoning potential respondents. They were asked if they had received the letter to refresh the memory of the respondent and establish the credentials of the caller. A convenient time for the interview was arranged for those who agreed to participate. Interviews usually occurred within five days of initial telephone contact. The proportion of potential respondents missing appointments rose appreciably as the scheduled time exceeded five days beyond telephone contact. A time lapse of seven days from posting of the letter to telephone contact was usually appropriate. Potential respondents had sufficient time to discuss the letter with family members yet not so long as to forget the contents of the letter.

Two trained interviewers were employed to minimize inconsistency of interviews. Close liaison was maintained between the interviewers throughout the survey process so that procedural problems were solved as they arose. Interview time ranged from fifteen minutes to two hours and averaged approximately one-half hour. About five interviews per interviewer were completed per day.

PERSONAL INTERVIEWS

At the time of the scheduled interview, an introductory statement established the credentials of the interviewer and the purpose of the survey. This was an informal paraphrase of the following:

> Hello, I'm _____ of Colorado State University in Fort Collins (hand respondent a copy of the introductory letter). A short time ago you received a similar letter in the mail. An appointment was made with you for this hour. This letter, from the Chairman of the Department of Economics, briefly explains the purpose of the study and the importance of talking with people in households like yours throughout the (Denver) (Fort Collins) area. I want to find out how you feel about water quality. I am interested in your enjoyment from using and viewing rivers and lakes (fish, waterfowl, water plants, and the water itself), and your satisfaction in knowing such natural environments are preserved.

Then the respondent was handed a copy of the questionnaire shown in Appendix B and asked to read along with the interviewer. Any questions raised by the respondent were clarified by the interviewer. First, the respondent was asked to provide socioeconomic information such as age, place of former residence, and income. Respondents were then asked their opinions about general environmental problems and perceived water quality in the South Platte River Basin.

Immediately after the respondents were asked to rate the quality of waterways in the basin (Question 11), they were shown a map illustrating the area encompassing the basin. Thus, they were familiar with the rivers and boundaries of the basin. When such information was needed during the remainder of the interview to provide an appropriate answer, respondents could review the map.

Next the respondents were read the following introduction to water quality and associated problems in Colorado:

> Coal development along with expanding mining operations may have significant effects on the quality of Colorado's water in the near future. As an aid in planning for the future I would like to find out how you feel about clean water for recreational activities. I have some questions which consider different ways of financing improved water quality. Let us consider three levels of water quality in a waterway such as the South Platte River Basin.

At this point respondents were shown pictures of three levels of water quality (see black and white examples in Figure 4.1). The interviewers pointed out the salient features of water pollution in each set of photographs. The six color photographs depict three representative levels of water quality in the South Platte River Basin. The respondents were told that in Situation A the water was in near natural condition with only trace pollutants. Situation B showed considerable reduction in water pollution but a greenish tint resulting from the presence of copper and other metals remained clearly visible. Situation C showed severe water pollution with a heavy load of mining waste.

The photographs were taken at three rivers located along the Colorado Front Range. Composition of the photographs was held constant insofar as possible. Site C with the most pollution was located on California Gulch near Leadville. Although California Gulch is in the Arkansas River Basin drainage system, pollution at the site was equal to Clear Creek in the South Platte River Basin.[3] By using this site, discolored water from mine drainage could be clearly shown. Site B was a case of intermediate pollution located directly below the Henderson molybdenum mine near Berthoud Falls. The site was on the West Fork of Clear Creek. Site A, the cleanest of the three sites, was located on the Poudre River above Fort Collins.

Color photographs were realistic in depicting evidence of visual pollution such as algae, weeds, and heavy metal discoloration but did not show nonvisual evidence of pollution such as odor and the presence of chemicals or toxic bacteria. Therefore, technical water quality data on the content of heavy metals at each site also was shown to the respondent along with a brief explanation of toxic effects on the human and wildlife community.

Water quality data for the three photo sites are shown in Table 4.3. Metals associated with acid mine drainage are listed with the amount of each in micrograms per liter. An asterisk (*) indicates that the concentration exceeds recommended drinking water standards. Recommended biological limits for fish and wildlife communities are also shown. Site C exceeds metal concentration recommended for drinking water. Site B exceeds total metal concentrations recommended for fish and wildlife communities. No heavy metals were measurable at Site A by the sampling and analytical methods used.

SIMULATED MARKET PAYMENT VEHICLES

Benefits of water quality were estimated using two simulated market payment vehicles: general sales taxes and residential water service fees. The two methods of payment were chosen to maximize the realism and credibility of the hypothetical market situation posited. Both approaches represent established routinized methods of paying for public services. Therefore, it was not difficult for most respondents to understand and accept the possibility of financing water pollution abatement by either method.

Residents of both cities were familiar with the practice of paying sales taxes. People were aware that revenue collected through sales taxes is used to provide public services. Participants in the survey could readily conceive of a public agency collecting a sales

52

Situation A   Situation C

Situation B

Figure 4.1. Photographs of Three Stream Sites, South Platte River Basin, Colorado, 1976.

Table 4.3. Heavy Metal Pollution at the Three Photograph Sites, South Platte River Basin, Colorado, 1973.

	Metal Concentration (Micrograms/Liter)			
Heavy Metals	C	B	A	Recommended Biological Limits
Arsenic	16*[a]	-	-	1,000
Cadmium	620*	<10	-	10
Copper	2,000*	10	-	10-20
Iron	50,000*	380*	-	300
Lead	450*	<50	-	5-10
Magnesium	66*	1.4	-	-
Manganese	28,000	580	-	1,000
Molybdenum	-	1	-	-
Nickel	50	<25	-	50
Selenium	48*	-	-	1,000
Vanadium	-	3.1	-	-
Zinc	100,000	90	-	30-70
Total Metals	181,250	<1,158.2	-	500

Source: Robert E. Moran and Dennis A. Wentz, Effects of Metal-Mine Drainage on Water Quality in Selected Areas of Colorado, 1972-1973, Colorado Water Resources Circular No. 25, (Denver: Colorado Water Conservation Board, 1974), 119 and 237, and Dennis A. Wentz, Effect of Mine Drainage on the Quality of Streams in Colorado, 1971-1972, Colorado Water Resources Circular No. 21, (Denver: Colorado Water Conservation Board, 1974), 42.

[a]An asterisk indicates this metallic concentration exceeds recommended drinking water standards.

tax and using the revenue to finance improved water quality. Respondents were asked to assume that a sales tax would be collected on all purchases in the South Platte River Basin for purposes of financing improved water quality in the river basin. This provision was designed to avoid the free-rider problem.

Residents of both cities also were familiar with the practice of paying for wastewater treatment through monthly water service fees. Homeowners and renters realized that revenues collected through water service charges provided treatment services. Most people could readily comprehend that a reduction in water pollution may raise the cost of operating treatment facilities. Paying these additional costs through increases in water service charges was a logical extension. Residents were asked to assume that a monthly assessment would be collected from all homeowners and apartment dwellers to finance improved water quality in the river basin. This provision was also designed to minimize the free-rider problem.

Respondents were instructed to assume that the methods of payment posited were the only ways to finance water quality improvement. This stipulation was designed to minimize the incidence of nonresponse and zero bids as protests against the particular method of payment. If a respondent was unwilling to pay anything, he was asked a series of questions to determine the reason. A respondent reporting that he did not consider his household harmed by water pollution and saw no reason to pay for improved water quality was recorded as having zero values and included in the sample. Respondents were considered to have refused financing water quality by the sales tax and fee methods if the response to a value question was zero and this reply was based on the belief that: (1) taxes were already too high, or (2) it was unfair to expect those adversely affected by water pollution to pay the costs of improvement, or (3) other reasons such as a lack of confidence in government entities to achieve the specified improvement.

Table 4.4 shows a comparison of the rate of acceptable responses to the water quality questions in the two cities. Only 10 percent of the sample of households refused to give an acceptable answer to at least one of the water quality value questions. It was appropriate for nonrecreationists not to answer the recreation use and option value questions. With respect to the sales tax method of payment, the higher rejection rate in Denver may have resulted from the fact that sales taxes were 1.5 cents per dollar higher in Denver than in Fort Collins. With respect to the water service fee method of payment, the higher rejection rate in Fort Collins may have resulted from resentment of recent increases in water service fees to accommodate rapid population growth.

Table 4.4 also shows the number of respondents willing to pay some positive amount of money for water quality in the two cities. Comparing the number who responded in each category of water quality value to the number willing to pay some positive amount reveals the number of acceptable zero responses. The proportion of respondents reporting acceptable zero value of recreation use, existence, and bequest demand for water quality was generally low, ranging from 2 to 8 percent. However, 25 to 28 percent of respondents reported acceptable zero values for option demand.

Immediately before asking respondents the contingent value questions, they were shown Table 4.5. This table indicates the annual amount of money in dollars that families in Colorado pay in sales taxes as estimated by the Internal Revenue Service in 1975. Internal Revenue Service income brackets are listed along the left edge of the table and family size is shown across the top. The first dollar value shown in columns of the table is the amount of sales tax a family would pay annually given a 5 percent sales tax rate. This is the state and city tax collected per dollar in Fort Collins at the time of the survey. Denver residents paid 6.5 percent in sales tax, but other metropolitan cities paid varying amounts depending on the particular suburb. The value in parentheses is the additional annual amount of money that would be paid with a one-quarter cent increment in sales tax. The annual amount paid in sales tax was then calculated for Denver area residents. Thus, respondents knew approximately how much money they paid in sales taxes and how much additional money would be paid for every one-quarter cent increment in sales tax before the contingent value questions were asked.

Table 4.4. Total Number of Residents Interviewed Compared to the Number Answering Water Quality Valuation Questions and the Number Willing to Pay in the South Platte River Basin, Colorado, 1976.

Response	Denver Metropolitan Area		Fort Collins		South Platte River Basin	
	Number	Percent	Number	Percent	Number	Percent
Total Number Interviewed	101	100.0	101	100.0	202	100.0
Number Answering Water Value Questions						
Recreation Value						
Sales Tax	85	84.2	89	88.1	174	86.1
Water Bill	82	81.2	78	77.2	160	79.2
Option Value						
Sales Tax	88	87.1	89	88.1	177	87.6
Water Bill	83	82.2	78	77.2	161	79.7
Existence Value						
Sales Tax	88	87.1	91	90.1	179	88.6
Water Bill	84	83.2	79	78.2	163	80.7
Bequest Value						
Sales Tax	88	87.1	93	92.1	181	89.6
Water Bill	84	83.2	80	79.2	164	81.2
Number Willing to Pay Some Amount						
Recreation Value						
Sales Tax	84	83.2	86	85.1	170	84.2
Water Bill	81	80.2	75	74.3	156	77.2
Option Value						
Sales Tax	65	64.3	58	57.4	123	60.9
Water Bill	59	58.4	52	51.5	111	55.0
Existence Value						
Sales Tax	85	84.2	86	85.1	171	84.7
Water Bill	77	76.2	74	73.3	151	74.8

Table 4.4. (continued)

Response	Denver Metropolitan Area		Fort Collins		South Platte River Basin	
	Number	Percent	Number	Percent	Number	Percent
Bequest Value						
Sales Tax	83	82.2	86	85.1	169	83.7
Water Bill	76	75.2	72	71.3	148	73.3

Table 4.5. Five Percent and One-Quarter Percent Incremental Sales Tax Value Estimates for Colorado Residents by Income and Family Size, 1975.

Income (Dollars)	Family Size					
	1	2	3	4	5	Over 5
Under 3,000	58 (2.92)	80 (4.00)	82 (4.08)	98 (4.92)	98 (4.92)	100 (5.00)
3,000- 3,999	73 (3.67)	97 (4.83)	102 (5.08)	118 (5.92)	122 (6.08)	125 (6.25)
4,000- 4,999	85 (4.25)	113 (5.67)	120 (6.00)	137 (6.83)	142 (7.08)	147 (7.33)
5,000- 5,999	97 (4.83)	127 (6.33)	137 (6.83)	152 (7.58)	162 (8.08)	167 (8.33)
6,000- 6,999	107 (5.33)	140 (7.00)	152 (7.58)	167 (8.33)	178 (8.92)	187 (9.33)
7,000- 7,999	117 (5.83)	153 (7.67)	167 (8.33)	182 (9.08)	195 (9.75)	205 (10.25)
8,000- 8,999	127 (6.33)	165 (8.25)	180 (9.00)	195 (9.75)	212 (10.58)	222 (11.08)
9,000- 9,999	135 (6.75)	177 (8.83)	193 (9.67)	207 (10.33)	227 (11.33)	238 (11.92)
10,000-10,999	143 (7.17)	187 (9.33)	207 (10.33)	218 (10.92)	242 (12.08)	255 (12.75)
11,000-11,999	152 (7.58)	197 (9.83)	220 (11.00)	230 (11.50)	257 (12.83)	272 (13.58)
12,000-12,999	160 (8.00)	207 (10.33)	232 (11.58)	242 (12.00)	270 (13.50)	287 (14.33)
13,000-13,999	168 (8.42)	217 (10.83)	243 (12.17)	252 (12.58)	283 (14.17)	302 (15.08)
14,000-14,999	177 (8.83)	227 (11.33)	255 (12.75)	262 (13.08)	297 (14.83)	317 (15.83)
15,000-15,999	183 (9.17)	235 (11.75)	267 (13.33)	272 (13.58)	310 (15.50)	330 (16.50)
16,000-16,999	190 (9.50)	243 (12.17)	278 (13.92)	282 (14.08)	323 (16.17)	343 (17.17)
17,000-17,999	197 (9.83)	252 (12.58)	288 (14.42)	292 (14.41)	335 (16.75)	357 (17.83)
18,000-18,999	203 (10.17)	260 (13.00)	298 (14.92)	302 (15.08)	347 (17.33)	370 (18.50)
19,000-19,999	210 (10.50)	268 (13.42)	308 (15.42)	310 (15.50)	358 (17.92)	383 (19.17)
20,000-20,999	214 (10.71)	274 (13.68)	314 (15.72)	316 (15.81)	365 (18.27)	391 (19.55)
21,000-21,999	218 (10.92)	279 (13.94)	320 (16.02)	322 (16.12)	372 (18.62)	398 (19.92)
22,000-22,999	223 (11.13)	284 (14.20)	326 (16.32)	329 (16.43)	379 (18.97)	406 (20.30)
23,000-23,999	227 (11.34)	289 (14.47)	333 (16.63)	335 (16.74)	387 (19.33)	414 (20.68)
24,000-24,999	231 (11.55)	295 (14.74)	339 (16.94)	341 (17.05)	394 (19.69)	421 (21.06)
25,000-25,999	235 (11.76)	300 (15.00)	345 (17.25)	347 (17.36)	401 (20.05)	429 (21.45)
26,000-26,999	239 (11.97)	306 (15.28)	351 (17.56)	353 (17.67)	408 (20.41)	437 (21.83)
27,000-27,999	244 (12.18)	311 (15.54)	357 (17.86)	360 (17.98)	415 (20.76)	444 (22.21)
28,000-28,999	248 (12.39)	316 (15.81)	363 (18.17)	366 (18.29)	422 (21.12)	452 (22.60)

Table 4.5. (continued)

Income (Dollars)	\multicolumn{5}{c}{Family Size}				
	1	2	3	4	5	Over 5
29,000-29,999	252 (12.60)	322 (16.08)	370 (18.48)	372 (18.60)	430 (21.48)	460 (22.98)
30,000-30,999	256 (12.81)	327 (16.35)	376 (18.79)	378 (18.91)	437 (21.84)	467 (23.36)
31,000-31,999	260 (13.02)	332 (16.62)	382 (19.10)	384 (19.22)	444 (22.20)	475 (23.75)
32,000-32,999	265 (13.23)	338 (16.88)	388 (19.40)	391 (19.53)	451 (22.55)	483 (24.13)
33,000-33,999	269 (13.44)	343 (17.15)	394 (19.71)	397 (19.84)	458 (22.91)	490 (24.51)
34,000-34,999	273 (13.65)	348 (17.42)	400 (20.02)	403 (20.15)	465 (23.27)	498 (24.89)
35,000-35,999	277 (13.86)	354 (17.69)	407 (20.33)	409 (20.46)	473 (23.63)	506 (25.28)
36,000-36,999	281 (14.07)	359 (17.96)	413 (20.64)	415 (20.77)	480 (23.99)	513 (25.66)
37,000-37,999	286 (14.28)	364 (18.22)	419 (20.95)	422 (21.08)	487 (24.34)	521 (26.04)
38,000-38,999	290 (14.49)	370 (18.49)	425 (21.25)	428 (21.39)	494 (24.70)	529 (26.43)
39,000-39,999	294 (14.70)	375 (18.76)	431 (21.56)	434 (21.70)	501 (25.06)	536 (26.81)

Source: U.S. Internal Revenue Service, 1975 Optional Sales Tax Tables, 1975 U.S. Individual Income Tax Return, (Washington, D.C.: Government Printing Office, 1974), p. 183.

ITERATIVE CONTINGENT VALUE APPROACH

The questions were designed to determine changes in willingness to pay for recreation use, option, existence, and bequest demand contingent on changes in water quality. Respondents were asked by how much they would be willing to increase their current sales taxes in cents per dollar and water service fee in dollars per month to obtain water quality in the South Platte River Basin. The starting point of the sales tax iterative valuation procedure was one-half cent with one-fourth cent incremental changes upward or downward. The starting point of the water service fee iterative valuation procedure was fifty cents per month with increments of fifty cents per month.

For each simulated market situation, respondents were asked to consider Situation C, with the largest amount of water pollution, as the beginning condition. An iterative procedure was used to discover the maximum amount of money with the respondent was willing to pay to improve water quality to intermediate Situation B and the pure water of Situation A. Respondents answered "yes" or "no" to questions expressed in the following form:

> Would you be willing to add one-fourth cent on the dollar to present sales taxes every year, if that resulted in an improvement from Situation C to Situation B?[4]

A "yes" response would lead the interviewer to raise the amount by one-fourth cent and repeat the question until a "no" answer was given. A "no" response resulted in a bid reduction until a "yes" answer was provided. The increment which resulted in the highest "yes" answer was recorded as the amount the respondent was willing to pay.

The question concerning willingness to pay additional sales taxes for enhanced enjoyment from recreation use was introduced with the following statement:

> Suppose a sales tax was collected from the citizens of the South Platte River Basin for the purpose of financing water quality in this basin. All of the additional tax would be used for water quality improvements to enhance recreational enjoyment. Every basin resident would pay the tax. All bodies of water in the river basin would be cleaned up by 1983. Assume that this is the only way to finance water quality improvement.

The questions concerning willingness to pay additional water service fees for enhanced enjoyment from recreation use was introduced with the following statement:

> Now let's consider a different way of financing water quality improvement. Suppose an extra water bill charge was collected from citizens of the South Platte River Basin for the purpose of financing water quality in this basin. All of the additional charge would be used for water quality improvements to enhance recreational enjoyment.

> Every basin resident would pay the charge. All bodies of
> water in the river basin would be cleaned up by 1983. Now
> assume that this is the only way to finance water quality
> improvement.
>
> Do you think it would be reasonable to add 50¢ to your
> water bill every month if that resulted in an improvement
> from Situation C to Situation B?

Respondents also were asked the value of improving water quality from Situation C to Situation A.

The year 1983 was specified because implementation of the Federal Water Pollution Control Act amendments of 1972 provided that abatement standards would be met by that year. The law called for industrial polluters to adopt the best available treatment technologies and municipal treatment systems to adopt the best practicable waste treatment technology by 1983. The national goal was to provide water quality suitable for fishlife and human contact recreation by that year. It was assumed that the payment of sales taxes and water service fees would become institutionalized and revenues collected after 1983 would be used to maintain water quality indefinitely in the future. This was reasonable since benefits from water quality improvement also would be realized indefinitely in the future.

The definition of recreation enjoyment was purposely left open to allow each respondent to provide his individual meaning. This approach was adopted so that respondents would be most likely to estimate their total benefits. Any particular definition of water-based recreation provided by the interviewer might have omitted an activity for which the respondent would be willing to pay. Thus, the definition included all water-based recreation activities such as: swimming, boating, fishing, and waterfowl hunting as well as noncontact recreation activities such as picnicking, camping, hiking, and sightseeing near water with enhanced aesthetic satisfaction of recreation experiences.

The questions concerning willingness to pay for option demand were designed to be as realistic as possible. An introductory statement explained the potential development of mining and the probable irreversible consequences to water quality in the river basin. The two alternative uses of waterways in the river basin were described, and substitution possibilities were minimized. Additional explanation was provided where necessary. The option value questions were as follows:

> In the near future, one of two alternatives is likely to
> occur in the South Platte River Basin. The <u>first alterna-
> tive</u> is that a large expansion in mining development will
> soon take place, creating jobs and income for the region.
> As a consequence, however, many lakes and streams would
> become severely polluted. It is highly unlikely, as is
> shown in Situation C, that these waterways could ever be
> returned to their natural condition. They could not be
> used for recreation. Growing demand could cause all other

waterways in the area to be crowded with other recreationists.

The second possible alternative is to postpone any decision to expand mining activities which would irreversibly pollute these waterways. During this time, they would be preserved at level A for your recreational use. Furthermore, information would become available enabling you to make a decision with near certainty in the future, as to whether it is more beneficial to you to preserve the waterways at level A for your recreation use or to permit mining development. Of course, if the first alternative takes place, you could not make this future choice since the waterways would be irreversibly polluted.

Given your chance of future recreational use, would you be willing to add ____ cents on the dollar to present sales taxes every year to postpone mining development? This postponement would permit information to become available enabling you to make a decision with near certainty in the future as to which option (recreational use or mining development) would be most beneficial to you. Would it be reasonable to add ____ to your water bill every month for this postponement?

The questions concerning willingness to pay for existence and bequest demands were introduced with a question concerning the chances of future recreation use of water in the river basin. This was designed to provide a subsample of nonuser values. The hypothesis to be tested was that nonusers as well as recreational users place a positive value on the preservation of water quality in the South Platte River Basin. The hypothetical situation presented to respondents also was prefaced with the condition that it is certain the respondent will not use the river basin for water-based recreation activities. The existence and bequest questions read to respondents were as follows:

(1) What would you estimate are the chances in 100 that you will travel to lakes and streams in the South Platte River Basin in the next year, for water-based recreation if they are preserved at level A? Do you anticipate any significant change in your chances for future years? (If "yes") What change?

(2) If it were certain you would not use the South Platte River Basin for water-based recreation, would you be willing to add ____ cents on the dollar to present sales taxes every year, just to know clean water exists at level A as a natural habitat for plants, fish, wildlife, etc.? Would it be reasonable to add ____ to your water bill every month for this knowledge?

(3) If it were certain you would not use the South Platte

River Basin for water-based recreation, would you be willing to add ____ cents on the dollar to present sales taxes every year to ensure that future generations will be able to enjoy clean water at level A? Would it be reasonable to add ____ to your water bill every month for this knowledge?

NOTES

1. Mountain Bell Telephone Company estimate, July 1977.
2. Phillip A. Meyer, Recreational and Preservation Values Associated with the Salmon of the Fraser River, Information Report Series No. PAC/IN-74-1, (Vancouver, B.C.: Environment Canada, Fisheries and Marine Service, Southern Operations Branch, Pacific Region, 1974).
3. It has been estimated that a 90 percent reduction in heavy metal pollution would be needed before fish could survive in Clear Creek. See Colorado Department of Health, Colorado Water Quality Report, 1975, (Denver: Colorado Department of Health, April 15, 1975), 24.
4. These questions were designed using the price-compensating measure of consumer surplus. This approach measures the amount of money an individual would be willing to pay above what he is already paying to continue consuming a commodity rather than forego consumption. The basic assumption of this measure is that "value in use" is greater than "value in exchange" for nonmarginal recreationists. An individual with an income of $$Y_0$ in the figure below would maximize utility on indifference curve I_1, at point A, if good X, (recreation value of water quality improvement) was unavailable or not purchased. If X were to be supplied the consumer would maximize utility on the higher indifference curve I_2 at point B, given budget slopes indicated by B_1 and B_2. The slopes of B_1 and B_2 will yield the price of X. The amount CV could be extracted from the individual's income and he would remain on the same indifference curve as if the good were not supplied. CV is the monetary approximation of the price compensating variation measure of consumer surplus.

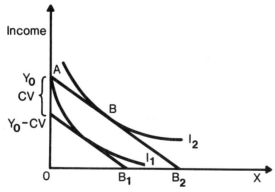

Figure 4.2. Price Compensating Variation of Consumer Surplus.

5
Option, Preservation, and Recreation Benefit Analysis

This chapter presents the results of the analysis of respondent's estimated benefits from improved water quality in the South Platte River Basin. First, average household benefits are presented for option and recreation use value of water quality to households expecting to continue using lakes and streams in the river basin for fishing, boating, swimming, and noncontact recreation activities such as picnicking and sightseeing near waterways. This is followed by a comparison of the preservation benefits of user and nonuser households, including existence and bequest values.

Average household benefits are presented for representative subsamples of residents in Denver and Fort Collins, and for the weighted average household in the river basin. Then, total annual benefits are estimated by extrapolation to all households living in the river basin. These estimates are used to calculate the present value of annual benefits. This illustrates the level of magnitude of benefits which could be used in benefit cost analysis of water pollution control in the river basin.

Subsequent sections of this chapter present several tests of the sensitivity of the results to changes in important variables. These include the effects of the payment vehicle on the benefit estimate, size of the geographic area in which water quality is improved, the timing of the water pollution control program, the achievement of an intermediate level of water quality improvement, and the size of city in the river basin. Standard statistical procedures are employed to test for significant difference between benefits reported by Denver and Fort Collins residents.

CALCULATION OF RIVER BASIN BENEFITS

The benefits of water quality were originally estimated for 1976. In order to update these values, the original estimates were increased by the percentage change in the Gross National Product Implicit Price Deflator[1] from the second quarter of 1976 through the first quarter of 1981. This index shows the difference in the cost of purchasing current year output valued at base period prices and at current market prices. The Implicit Price Deflator was 133.06 during the second quarter of 1976, the period during which the survey was taken. The index stood at 187.30 for the first quarter of 1981. Thus, the prices of goods and services included in the 1981 GNP increased 40.76 percent over second quarter 1976 prices. Accordingly, our original 1976 values are increased by 40.76 percent in the tables in this chapter. These water quality benefit estimates for 1981 are likely to be conservative. Per capita personal income in the river basin has increased by approximately 12 percent per year over the 5 year period, or somewhat more rapidly than the price index. Benefits would be expected to increase proportionally more than real income.

South Platte River Basin benefits were estimated from Denver and Fort Collins survey data by weighting them proportionately by their representative basin populations and summing the two values. The Denver metropolitan area population of 1,267,000 persons, excluding Boulder County, was 72.7 percent of the 1,742,900 persons in the South Platte River Basin, Colorado, in 1976. Fort Collins estimates were weighted by the remaining 27.3 percent of the basin population.

By 1980, the South Platte River Basin population had increased approximately 13.3 percent to 1,974,868. The Denver Standard Metropolitan Statistical Area (SMSA) population, excluding Boulder County, was estimated at 1,430,296 residents, or about the same proportion of the total basin population as in 1976. The weights used to estimate the river basin benefits were therefore left unchanged in deriving the 1981 estimates.

OPTION VALUES

Option value was defined in this study as household willingness to pay for the option to choose to engage in water-based recreation activities in the future given that imminent expansion of mineral and energy development in the river basin could preclude such activity without payment of the fee. The protection of water quality provides the option to make a future decision between two alternative uses of waterways in the river basin, either for water-based recreation or for wastewater discharge from industrial, mineral, and energy development, under conditions of near certain knowledge about which use will be more beneficial. The economic significance of option value and other preservation values is that they shift the vertical intercept of the demand curve for water quality.

Option value constitutes a substantial part of the total value of water quality in the South Platte River Basin, as shown in Table 5.1. Option value added about 40 percent to the recreation benefits from enhanced enjoyment of water-based recreation. River basin resident households reported willingness to pay an average of nearly $32 annually in sales taxes to improve water quality in the river basin for option demand. The 95 percent confidence interval around this average value ranged from $22 to $42. This was the value reported by 81 percent of the sample households stating a positive probability of future water-based recreation in the river basin. Approximately 20 percent of respondents answering the option value question reported zero benefits.

This is believed to be the first empirical test of the concept of option value for any nonmarket good. Thus, there is very little empirical literature available with which to compare the results. Recently, Brookshire, Eubanks, and Randall[2] used the contingent value approach to estimate option value of $22 annually for grizzly bear hunting in Wyoming. It was specified that revenues from a special stamp would be used to provide suitable habitat for bear, currently protected by a hunting moratorium. If and when the moratorium is lifted only those paying for the option would be allowed to hunt bear. In the same study, option value of $30 annually was reported for the option of purchasing a big horn sheep hunting license when demand exceeds supply.

Low[3] used the contribution of time, money, and services by members of Alaska conservation organizations in support of the Alaska Lands Bill to estimate willingness to pay for the wilderness option. Members were willing to pay an average of $218 to $846 per year. The value of time in the lower estimate was based on the wage value of services provided and the higher estimate was based on the income of the donor. Members of special interest groups were willing to pay substantially more in option demand for wilderness at least in support of a campaign of a few years duration than the general public would pay annually in the long run. Walsh, Gillman, and Loomis[4] used the contingent value approach to estimate option value of about $10 per household annually from designating all potential wilderness areas in Colorado. Acreage of potential wilderness designation differed substantially with 10 million acres in Colorado compared to 50 to 100 million acres in Alaska.

RECREATION USE VALUES

South Platte River Basin resident households reported willingness to pay an average of $80 annually in sales taxes to improve water quality in the river basin for recreation use (Table 5.1). The 95 percent confidence interval around this average value ranged from $65 to $95. This means that there is only one chance in twenty that the findings would vary by more than this amount.

This was the value reported by 81 percent of the sample households who expect to continue to use lakes and streams in the river basin for fishing, boating, swimming, and noncontact recreation activities such as picnicking and sightseeing near water with enhanced aesthetic satisfaction of such recreation experiences. These resident households reported an average of about fifteen water-based recreation activity days annually in the river basin, so that the recreation use value of improved water quality was equivalent to an average of $5.30 per activity day. The 95 percent confidence interval around the mean value ranged from $4.30 to $6.30 per activity day.

The recreation benefits of water quality improvement to residents of the river basin were somewhat higher than benefits to tourists in Rocky Mountain National Park.[5] A substantial portion of the national park is located within the South Platte River Basin. In the summer of 1973, park visitors reported they were willing to pay an entrance fee of $6 per household day to avoid water pollution where they engage in water-based recreation activities. Adjusted for the 77.2 percent inflation between 1973 and the first quarter of

Table 5.1. Resident Household Mean Willingness to Pay Additional Sales Taxes and Water Service Assessments for Water Quality in the South Platte River Basin, Colorado (1981 dollars)

Water Quality Improvement Values	Water Fee Per Month (dollars)	Tax Rate (cents)	Annual Dollars — Water Fee	Annual Dollars — Sales Tax[a]
Denver Metropolitan Area				
Option Value	$0.70	0.54¢	$8.45	$25.77
95% Confidence Interval			(6.45-10.57)	(18.98-32.88)
Percent of Total Value	17.7%	15.5%	17.7%	16.5%
Number Reporting	83	88		
Existence Value[b]	$0.76	0.98¢	$9.12	$36.64
95% Confidence Interval			(6.73-11.36)	(2.51-70.79)
Percent of Total Value	19.1%	27.9%	19.1%	23.5%
Number Reporting	14	15		
Bequest Value[c]	$0.64	0.59¢	$7.77	$23.13
95% Confidence Interval			(5.53-10.16)	(9.39-36.87)
Percent of Total Value	16.3%	16.7%	16.3%	14.8%
Number Reporting	14	15		
Recreation Value[d]	$1.86	1.41¢	$22.30	$70.63
95% Confidence Interval			(18.28-26.31)	(60.44-80.84)
Percent of Total Value	46.8%	39.8%	46.8%	45.2%
Number Reporting	82	85		
Total Preservation and Recreation Value	$3.97	3.53¢	$47.64	$156.17
Fort Collins				
Option Value	$1.41	1.19¢	$16.89	$47.93
95% Confidence Interval			(8.00-25.53)	(28.31-67.55)
Percent of Total Value	24.0%	25.1%	24.0%	22.9%
Number Reporting	78	89		
Existence Value[b]	$0.81	0.70¢	$9.80	$31.21
95% Confidence Interval			(7.56-12.15)	(7.50-69.90)

Percent of Total Value	13.9%	14.8%	13.9%	14.9%
Number Reporting	9	9		9
Bequest Value[c]	$0.59	0.56¢	$7.09	$25.93
95% Confidence Interval			(3.80-10.29)	(-5.75-65.50)
Percent of Total Value	10.1%	11.8%	10.1%	12.4%
Number Reporting	9	9		9
Recreation Value[d]	$3.04	2.29¢	$36.48	$104.16
95% Confidence Interval			(16.60-56.39)	(71.89-136.45)
Percent of Total Value	51.9%	48.2%	51.9%	49.8%
Number Reporting	89	89		
Total Preservation and Recreation Value	$5.86	4.76¢	$70.26	$209.23

Aggregated South Platte River Basin

Option Value	$0.90	0.76¢	$10.81	$31.81
Percent of Total Value	20.1%	19.0%	20.1%	18.6%
Number Reporting	161	177		
Grouped t-value[b]	1.01	1.51		
Existence Value[b]	$0.77	0.90¢	$9.29	$35.16
Percent of Total Value	17.2%	23.3%	17.2%	20.6%
Number Reporting	23	24		
Grouped t-value[e]	0.99	1.69		
Bequest Value[c]	$0.63	0.58¢	$7.60	$23.89
Percent of Total Value	14.1%	15.0%	14.1%	14.0%
Number Reporting	23	24		
Grouped t-value[e]	0.88	1.55		
Recreation Value[d]	$1.55	1.17¢	$26.18	$79.78
Percent of Total Value	48.6%	42.7%	48.6%	46.8%
Number Reporting	171	174		
Grouped t-value[e]	0.81	1.32		
Total Preservation and Recreation Value	$4.50	3.86¢	$53.88	$170.64

[a] U.S. Internal Revenue Service, 1975 U.S. Individual Income Tax Return, 1975 Optional State Sales Tax

Table 5.1. (continued)

Tables, (Washington, D.C.: Government Printing Office, 1974), 183.
[b]Willingness to pay estimate for a subsample of nonrecreationists for the benefit derived from the knowledge that a natural area exists as a habitat for various species of flora and fauna.
[c]Willingness to pay estimate for a subsample of nonrecreationists for the benefit derived from the assurance that future generations will have access to a natural environment in the South Platte River Basin, Colorado.
[d]Defined as any water-associated recreation benefit derived from improved water quality by 1983.
[e]Tests the statistical significance of the difference between grouped mean values. At the 5 percent level, there is no significant difference between the Denver and Fort Collins mean water quality value estimates.

1981, as measured by the GNP Implicit Price Deflator, and dividing by family size of three persons, tourist benefits would be $3.54 per recreation activity day. This would be somewhat below the 95 percent confidence interval around our estimate of benefits to residents of the South Platte River Basin. Water quality is higher in the park than in the river basin as a whole. The park is one of the unique natural areas of the nation, with pristine rivers and lakes, and majestic mountain peaks. This unique natural setting may have influenced the values estimated by respondents. The important finding is that tourists as well as residents are willing to pay for water quality in the South Platte River Basin.

The estimate of recreation benefits to residents of the South Platte River Basin was similar to benefits reported in a study of water quality in the Merrimack River Basin, located in New Hampshire and Massachusetts. Oster[6] used the contingent value approach in telephone interviews with 200 residents of the river basin. The author reported that average willingness to pay for pollution abatement was $12 per person annually in 1973 dollars. If family size was equal to the national average of three persons, this would be equivalent to a value of $36 per household living in the river basin. Adjusted for the 77.2 percent inflation from 1973 to 1981, benefits would be $64 per household annually. This would be on the low side of the 95 percent confidence interval around our estimate of benefits to residents of the South Platte River Basin.

Knetsch and Davis[7] reported that the contingent value approach to valuing forest recreation resulted in benefit estimates which were not significantly different from the travel cost approach. Our contingent value estimate of the recreation use benefits of improved water quality in the South Platte River Basin compares favorably with the findings of Bouwes and Schneider[8] who used the travel cost approach to estimate the recreation use benefit of water quality in a small lake located in southeastern Wisconsin. From the information available, it was possible to calculate the expected shift in the travel cost demand curve with severely polluted water in the lake, i.e., a change from 3 to 23 on the pollution index. Demand would decline by 3.8 trips per year from 13.9 trips with a pollution index of 3 to 10.1 trips with a maximum pollution index of 23. We estimate that recreation benefits measured as the consumer surplus area between demand curves with and without this level of pollution would be roughly $60 annually per household in 1976 dollars. Adjusted for inflation to 1981, benefits would be $85 per household annually. This would be well within the 95 percent confidence interval of our estimate for the South Platte River Basin.

The South Platte River Basin results are comparable to those reported in a recent estimate of national recreation benefits of water quality prepared for the President's Council on Environmental Quality. Freeman[9] reviewed the limited secondary data and concluded that national recreation benefits were most likely $6.7 billion annually with a range of $4.1-$14.1 billion in 1978 dollars. Adjusted for the 24.8 percent inflation from 1978 to 1981, as measured by the GNP Implicit Price Deflator, this benefit estimate would increase to $8.4 billion with a range of $5.1-$17.6 billion. The substantial range in estimates reflects the uncertainty associated with limited data, and should become more precise with the development of improved procedures and data.

With approximately 80 million U.S. households averaging about 2.85 persons in 1981, estimated national recreation use benefits extrapolated from our South Platte River Basin data would average $5.2 billion annually with a range of $4.2-$6.2 billion in 1981 dollars. National option benefits would average $2.1 billion with a range of $1.4-$2.7 billion. Thus, our estimated national recreation use and option benefits would total $7.2 billion annually with a range of $5.6-$8.9 billion in 1981 dollars. The range represents the 95 confidence interval around the mean value.

This extrapolation of our river basin data resulted in a national recreation benefit estimate which was somewhat lower and more conservative than Freeman's most likely case. Moreover, our 95 percent confidence interval provided a basis for estimation of a more precise range of estimate, well within Freeman's acceptable range. South Platte River Basin residents probably have experienced less water pollution than in the industrial centers of the nation. Also, Colorado residents may have more close substitutes available, i.e., pristine high mountain rivers and lakes elsewhere in the Rocky Mountains.

The annual value of water quality protection was similar in amount to the reported recreation value of air quality. A 1972 survey[10] of a representative sample of resident and tourist households in New Mexico reported average willingness to pay of $85 annually in sales taxes to avoid aesthetic damages from air pollution, transmission lines, and spoil banks associated with the coal-based power plant at Fruitland, New Mexico. The reliability of this study has been tested by replication under similar conditions. A 1974 survey[11] of a sample of resident households in Farmington, New Mexico, and recreation users of the Four Corners region of New Mexico and Arizona reported average willingness to pay of $82 annually to avoid visibility damages from air pollution. Recreation households were willing to pay $2.44 per day. In the summer of 1974, a representative sample of recreation households were interviewed while visiting Lake Powell in Glen Canyon National Recreation Area.[12] Recreation households reported average willingness to pay an entrance fee of $2.77 per day to avoid air pollution damage from the Navajo Power Plant visible south of the lake. While not designed as replications, these studies demonstrate reasonable consistency.

EXISTENCE AND BEQUEST VALUES

Preservation benefits of water quality in the South Platte River Basin were defined to include both the value placed on existence of a natural ecosystem and the value of its bequest to future generations. Table 5.2 compares the existence and bequest values reported by recreation user and nonuser households.

River basin households reported willingness to pay an average of $94 annually in sales taxes to improve water quality in the river basin for preservation demand. The 95 percent confidence interval around this mean value ranged from $69 to $119. Preservation benefits include average bequest value of $46 annually with a range of

$34-$59 and existence value of $48 annually with a range of $35-$60. These were the preservation values reported by the entire sample of households interviewed including 81 percent who expected to continue using lakes and streams in the river basin for fishing, boating, swimming, and noncontact recreation activities such as picnicking and sightseeing near waterways. If there were no water-based recreation activities in the South Platte River Basin, resident households would still value improved water quality by this amount. The estimates of preservation values by recreation users were premised on the assumption that the respondents knew with certainty they would not engage in water-based recreation activities in the river basin.

Since preservation values were defined as the benefits which would remain in the absence of recreation use, we believe that to avoid upward bias, the preservation values reported by users should not be added to their recreation use and option benefits. As a first approximation, preservation benefits of the 19 percent of respondents who reported a zero chance of future water-based recreation use of the river basin were extrapolated to all resident households. This procedure involves the premise that recreation user households would be willing to pay no more for preservation demands than nonuser households.

Nonuser households reported average total preservation values of $59 annually with a range of $12-$107. The range represents the 95 percent confidence interval around the mean value. This includes average bequest value of $24 annually with a range of $12-$36 and existence value of $35 annually with a range of $0-$71. These were the preservation values reported by the 19 percent of the sample stating a zero chance of future use of the river basin for water-based recreation activities. The two-tailed t-test of significance revealed that the value estimated for this small subsample of respondents was significantly different from zero in the case of existence and bequest values in Denver but not in Fort Collins. A one-tail test may be more appropriate for data such as this where only positive values were reported. Existence value in Fort Collins was significant in a one-tail test at the 95 percent confidence level, and bequest value was significant at the 90 percent level. Thus, we consider the average values reasonable and a larger subsample would increase their significance.

These benefit estimates were somewhat higher than a recent estimate of nonuse benefits prepared for the President's Council on Environmental Quality. Freeman[13] reviewed the scant evidence on preservation values and concluded that national nonuse benefits were most likely $2 billion annually with a range of $1-$5 billion in 1978 dollars. Adjusted for the 24.8 percent inflation from 1978 to 1981, as measured by the GNP Implicit Price Deflator, this benefit estimate would increase to $2.5 billion with a range of $1.3-$6.2 billion. The substantial range in estimate reflects the uncertainty associated with limited data, and should become more precise with the development of improved procedures and data.

With an estimated 80 million U.S. households averaging 2.85 persons in 1981, our estimate of national bequest benefits would average $1.9 billion annually with a range of $1.0-$2.9 billion in 1981 dollars. The range represents the 95 percent confidence interval around

Table 5.2. Resident Recreation User and Nonuser Willingness to Pay Additional Water Service Assessments and Sales Taxes to Improve Water Quality for Existence and Bequest Benefits in the South Platte River Basin, Colorado (mean values in 1981 dollars).

Water Quality Improvement Values	Water Fee Per Month (dollars)	Tax Rate (cents)	Annual Dollars Water Fee	Annual Dollars Sales Tax
Denver Metropolitan Area				
All Resident Households (assuming zero chance of future recreation use)				
Existence Value	$0.97	0.86¢	$11.65	$40.75
95% Confidence Interval			(9.64-13.54)	(31.91-49.58)
Number Reporting	84	88		
Bequest Value	$1.04	0.79¢	$12.50	$40.09
95% Confidence Interval			(9.92-14.96)	(31.95-48.22)
Number Reporting	84	88		
Total	$2.01	1.65¢	$24.15	$80.84
Resident Nonuser Households (reporting zero chance of future recreation use)				
Existence Value	$0.76	0.99¢	$9.12	$36.64
95% Confidence Interval			(6.73-11.36)	(2.51-70.79)
Number Reporting	14	15		
Bequest Value	$0.65	0.59¢	$7.77	$23.13
95% Confidence Interval			(5.53-10.16)	(9.39-36.87)
Number Reporting	14	15		
Total	$1.41	1.58¢	$16.89	$59.77
Fort Collins				
All Resident Households (assuming zero chance of future recreation use)				
Existence Value	$1.64	1.54¢	$19.59	$66.04
95% Confidence Interval			(11.34-27.77)	(45.04-87.05)
Number Reporting	79	91		
Bequest Value	$1.69	1.55¢	$20.27	$63.47
95% Confidence Interval			(11.43-29.01)	(42.48-84.47)
Number Reporting	80	93		

Total		$3.33	$129.51
Resident Nonuser Households (reporting zero chance of future recreation use)			
Existence Value	3.09¢	$39.86	
95% Confidence Interval	0.70¢	$9.80	$31.21
		(7.56-12.15)	(-7.50-69.90)
Number Reporting	9		
Bequest Value	0.56¢	$7.09	$25.93
95% Confidence Interval		(3.80-10.29)	(-5.75-65.50)
Number Reporting	9		
Total	1.26¢	$16.89	$57.14

South Platte River Basin

All Resident Households (assuming zero chance of future recreation use)			
Existence Value	1.04¢	$13.85	$47.66
Number Reporting	163		
Bequest Value	0.99¢	$14.70	$46.46
Number Reporting	164		
Total	2.03¢	$28.55	$94.12
Resident Nonuser Households (reporting zero chance of future recreation use)			
Existence Value	0.90¢	$9.29	$35.16
Number Reporting	24		
Bequest Value	0.58¢	$7.60	$23.89
Number Reporting	24		
Total	1.48¢	$16.89	$59.05

the mean value. Our estimate of national existence benefits would average $2.8 billion annually with a range of zero to $5.7 billion. Thus, our estimated national nonuse benefits would total $4.7 billion annually with a range of $0.9-$8.6 billion in 1981 dollars. This extrapolation of our data results in a higher national nonuse benefit estimate than Freeman's and one with a somewhat wider confidence interval.

The estimated preservation benefits of water quality in the South Platte River Basin were relatively small compared to estimates for the Fraser River Basin, British Columbia, Canada.[14] Preservation values were reported as $233 per household annually, which increased salmon fishing values by 54 percent. It should be noted that these values were reported for the preservation of a free flowing river system from impoundment by a large water development project, and are not strictly comparable to values from the preservation of water quality alone.

TOTAL ANNUAL BENEFITS PER HOUSEHOLD

The total annual benefits of water quality per household living in the river basin were estimated as $149. The 95 percent confidence interval around this average value ranged from $82 to $217. This included recreation benefits of $64 with a range of $52-$77, option value of $26 with a range of $18-$34, bequest value of $24 with a range of $12-$36, and existence value of $35 with a range of $0-$71. These weighted total annual benefits per household were estimated in two steps: (1) for 81 percent of the households who expected to be recreation users in the future, and (2) for 19 percent of the households who did not expect to be users in the future.

The total annual benefits of water quality to households engaging in water-based recreation and also possessing, option, existence, and bequest demands were estimated as $171 annually. This included recreation use benefits of $80, option value of $32, existence value of $35, and bequest value of $24. Existence and bequest values were those reported by a subsample of residents who currently do not expect to use the river basin for water-based recreation activities, on the assumption that they are representative of existence and bequest values of the general population of the river basin. This may be a conservative estimate, as the experience and appreciation gained in the recreation use of these resources may result in somewhat higher estimates of existence and bequest values than for the nonuser sample of the population.

The average total benefits to 19 percent of the households who reported they do not expect to make recreational use of the waterways in the river basin were estimated as $59 annually. This included existence value of $35 and bequest value of $24 to the relevant population of nonusers. Recreation use and option value estimates were excluded from nonuser benefits since they reported a zero probability of future recreation use in the river basin. These nonuser households represented 14.9 percent of the Fort Collins population and 20.8 percent of the Denver population.

These total benefit estimates are similar to reported average total benefits of water quality in the Charles River Basin,

Massachusetts. Gramlich[15] used the contingent value approach to estimate that resident households of the Boston area would be willing to pay an average of $30 per household in annual taxes to improve water quality in the Charles River Basin, Massachusetts (in 1973 dollars). To improve water quality in other river basins throughout the U.S., residents of Boston were willing to pay an additional $25 in annual taxes. Thus, total willingness to pay for improved water quality in the U.S. including the Charles River Basin was reported as $55 per year. The author did not ask respondents to allocate the stated willingness to pay among recreation use value and option or preservation value. Thus, the benefit estimate can be interpreted as being based on each respondent's perception of a combination of recreation use, option, and nonuser preservation values.

The improvement in Charles River water quality was defined as from the current relatively polluted level to a level suitable for swimming but not suitable as a public water supply which was ruled out for reasons of cost. At the time of the survey, water quality in the Charles River was not sufficient for swimming but was clean enough for wildlife use in the upper and middle parts of the 80 mile river. Water quality tended to decline with unpleasant odor and health hazard in the heavily-used lower basin area. We estimate that the perceived improvement in water quality would be equivalent to an increase from a rating of 75 to 25 on a 100-point scale with zero as clean enough for use as a public water supply and 100 a health hazard.

To compare the Charles River results to our study, water quality benefits were adjusted for inflation of 77.2 percent from 1973 to 1981, as measured by the GNP Implicit Price Deflator, and doubled for a 100 percent rather than a 50 percentage point improvement in water quality. On this basis, residents of the Charles River Basin were willing to pay an average of $106 per household annually for water quality in 1981. The 95 percent confidence limit around this mean value ranged from $76 to $140. While this was less than the average of $149 annually households living in the South Platte River Basin were willing to pay for water quality, it was within the 95 percent confidence interval of $82-$217. Moreover, Boston residents reported a willingness to pay an additional $88 per houschold annually to improve water quality throughout the U.S., while South Platte River Basin residents reported an unwillingness to pay for water quality in other river basins.

ANNUAL BENEFITS AND PRESENT VALUE OF FUTURE BENEFITS

Estimates of total annual benefits from improved water quality were prepared for the South Platte River Basin. These were used to calculate the present value of a future stream of annual benefits. Table 5.3 shows the annual and present values of benefits from recreation use, option, existence, and bequest demand. The method of payment in the discussion of benefits below is sales taxes unless otherwise stated.

Total annual benefits of water quality in the South Platte River Basin were estimated as $95.1 million in 1981. This included recreation use benefits of $33.2 million, option value of $18.8 million,

Table 5.3. Annual and Present Value of Water Quality in the South Platte River Basin, Colorado (1981 dollars).

Water Quality Values	Denver Metropolitan Area		Fort Collins		South Platte River Basin	
	Annual Value	Present Value[a]	Annual Value	Present Value[a]	Annual Value	Present Value[a]
Option Value						
Water Fee	$3,519,180	$47,717,803	$337,310	$4,573,695	$6,374,905	$86,439,390
Sales Tax	10,732,458	145,524,854	635,277	8,613,925	18,759,089	254,360,529
Existence Value						
Water Fee	4,795,716	65,026,658	195,716	2,653,776	6,780,344	91,936,869
Sales Tax	19,266,997	261,247,417	623,295	8,451,458	25,661,667	347,954,807
Bequest Value						
Water Fee	4,085,823	55,400,990	141,594	1,919,919	5,546,890	75,212,068
Sales Tax	12,162,818	164,919,566	517,848	7,021,668	17,436,212	236,523,214
Recreation Value						
Water Fee	9,287,303	125,929,532	728,542	9,878,536	10,903,211	147,840,149
Sales Tax	29,415,347	398,852,163	2,080,179	28,205,817	33,226,056	450,522,793
Total Preservation and Recreation Value						
Water Fee	21,688,022	294,074,875	1,403,162	19,025,925	29,605,350	401,428,475
Sales Tax	86,395,901	1,171,469,844	3,859,599	52,333,546	95,083,024	1,289,261,342

[a]Discounted at 7 3/8 percent, the recommended discount rate for water resource development projects by the U.S. Water Resources Council in 1981.

existence value of $25.7 million, and bequest value of $17.4 million. These total annual benefits were estimated in two steps: (1) for the 81 percent of the households who expect to be recreation users in the future, and (2) for the 19 percent of the households who expect to be nonusers in the future. There were an estimated 730,000 households in the river basin in 1981, with 525,846 in the Denver SMSA excluding Boulder County. The annual benefit estimates were a weighted average based on the proportion of basin population in the Denver metropolitan area and the proportion of the population in the nonmetropolitan areas. It was assumed that Fort Collins was representative of the nonmetropolitan areas of the river basin.

The present value of a perpetual stream of annual benefits from water quality in the South Platte River Basin was calculated as $1.3 billion. This included recreation use value of $451 million, option value of $254 million, existence value of $347 million, and bequest value of $236 million. Present value is the amount of money that would have to be invested at interest today in order to yield the specified annual benefits from improved water quality for an indefinite period of time. The formula is $PV = B/i$ where PV is the present value of a perpetual stream of annual benefits, B is the annual benefits from water quality improvement, and i is the Federal discount rate of 7 3/8 percent used in calculation of benefits and costs of public water resource projects in 1981.

The calculation of present value of future benefits is included for illustrative purposes, and is likely to be a low estimation for a number of reasons. Future benefits are assumed to remain constant at 1981 levels, which seems rather unlikely to occur. Population is expected to continue to grow rapidly in some parts of the river basin, as migration from other parts of the nation continues to occur. Also, the results of the socioeconomic regression analysis suggest that variables such as future growth in income, increased education levels, and changes in the age characteristics of the population will significantly increase willingness to pay for improved water quality. Substitute recreation areas may become crowded and polluted, and with more leisure time available the proportion of the population who engage in water-based recreation activities in the river basin may increase. These trends suggest that the present value of the benefit stream may prove conservative. In addition, tourists account for approximately 30 to 40 percent of the total water-based recreation activities in the river basin, and benefits to them were not estimated in this study. Other research suggests that including the value of improved water quality to nonresident tourists would increase resident values shown here by at least 28 percent.[16]

METHOD OF PAYMENT

Survey respondents indicated a greater willingness to pay for improved water quality when the method of hypothetical payment was an increase in sales taxes rather than an increase in water service fees. Table 5.1 shows that willingness to pay additional water service fees was about one-third as much as willingness to pay additional sales taxes. This was an unexpected result, as previous research suggested alternative methods of payment would not affect willingness to pay.

In a controlled test of the contingent value approach to estimation of willingness to pay, Bohm[17] found that several hypothetical methods of payment did not significantly affect resulting values as compared to actual payment.

The relationship between values obtained by the two methods of payment was nearly identical in both Denver and Fort Collins. For example, willingness to pay additional water fees for recreation use was reported as 32 percent as much as sales tax in Denver, and 35 percent as much in Fort Collins. This suggests that factors influencing the relative payment were general in nature. A sales tax is collected from everyone who purchases goods and services in the taxing district, including tourists, whereas water service fees are paid by property owners and only indirectly by renters. This is the free-rider problem in which tourists tend to escape payment when water quality is improved with water-sewer district revenues.[18] Respondents were more reluctant to participate in the water bill value estimation procedure. This may have resulted from perceived inequities. In portions of both cities which were not metered, small families were required to pay the same flat fee as large families. Also, with an average water bill of $10 to $15 per month, an incremental willingness to pay of 50 cents per month is a larger percentage of the total water bill than one-quarter cent in additional sales tax. Although the average willingness to pay additional sales tax for improved water quality amounted to more total annual dollars, it was approximately the same percentage of the annual sales taxes as the water fee estimates were of the annual water bill. Moreover, recent experience with escalating water-sewer fees also may have resulted in an understatement of willingness to pay for water quality.

LEVEL OF WATER QUALITY

Table 5.4 shows the relationship between the level of water pollution control and willingness to pay for enhanced recreation use. Respondents were not asked for the value of option, existence, and bequest demands related to an intermediate level of water quality. The average values suggest that recreation use benefits from water quality improvement increase at a decreasing rate, which is consistent with decreasing marginal utility of consumption observed for private consumption goods.

In Fort Collins, improving polluted water to an intermediate water quality level (from C to B photographs) accounted for 63 percent of total recreation use benefits from clean water. This is similar to research results concerning benefits of air quality improvement. It has been shown that improving air quality to an intermediate level accounted for 57 to 59 percent of total aesthetic benefits from clean air.[19]

A greater proportion of the benefits to residents of Denver are realized by improving water quality from the worst conditions of pollution to an intermediate level, from C to B. The intermediate level of water pollution control accounted for 74 percent or $52 annually of total sales tax values reported for improving water quality from polluted to clean levels. This is consistent with recent experience in the Denver metropolitan area. Benefits have accrued to residents from partial improvement of water quality from year to year. In central Denver, the quality of the South Platte River has been improved from a level which would not support fishlife to a level which now can sustain lower levels of fishlife such as catfish and bullheads. This may be considered an improvement from a classification of polluted to an intermediate level of water quality (from C to B), for the river does not yet contain sufficient dissolved oxygen to sustain game fish such as trout. When the highest level of improvement is eventually achieved, it may be considered an improvement from intermediate water quality to clean water (from B to A). When the South Platte River becomes clean, the additional recreation use benefits to Denver residents will increase by $18 annually, representing 26 percent of the total sales tax value of water quality reported by Denver residents.

Since an improvement in water quality from Situation C to Situation B accounts for a larger reduction in pollution than from B to A, these estimates appear to be reasonable. Heavy metal pollution was used as a proxy of water pollution rather than the more general indices of dissolved oxygen and biochemical oxygen demand. Heavy metal effluent results in many of the same problems as do other effluents, such as fish kills. The presence of some metals, such as iron which results in acid formation, have a more pronounced effect on water quality and the adjacent wildlife community. However, an objective basis for weighting was unavailable. Although Situations B and A are quite close together in metals content relative to Situation C, there is a significant change in the level of water quality. Fish and wildlife are still limited by the toxic effects of the water in Situation B while the water in Situation A is pure and nontoxic. Situation B is representative of areas with approximately 1,158 micrograms of heavy metals per liter of water, Situation C with approximately 181,250 micrograms of heavy metals per liter of water, and Situation A where undetected trace elements remain represents clean water. Situation C depicts the worst water quality level within the South Platte River Basin where metallic content exceeds recommended drinking water standards and biological limits for fish survival. Thus, respondents provided estimates of benefits based on their personal experience and judgment of the actual amount of pollution in the river basin.

EFFECTS OF DELAYED WATER QUALITY IMPROVEMENT

Estimates of willingness to pay for improved water quality discussed so far have been based on the premise that all bodies of water in the river basin would be cleaned up by 1983 and then maintained in a clean state indefinitely. Table 5.5 shows that if a postponement of environmental quality objectives resulted in a delay in improvement of water quality in the South Platte River Basin until the year 2000, the proportion of respondents willing to pay some amount of additional sales tax for improved water quality would decline by 11 percent. If it is not possible to improve water quality in the South Platte River Basin until the year 2000, annual willingness to pay sales taxes for recreation use would fall by an average of $15 per household or 18.5 percent. Table 5.5 shows the relative values

Table 5.4. Effect of Level of Pollution Abatement on Resident Household Willingness to Pay Additional Water Service Assessments and Sales Taxes to Improve Water Quality for Recreation Use in the South Platte River Basin, Colorado (mean values in 1981 dollars).

Water Quality Improvement Values	Water Fee Per Month (dollars)	Tax Rate (cents)	Annual Dollars	
			Water Fee	Sales Tax
		Denver Metropolitan Area		
Improvement from Severe to Moderate Pollution (C to B)	$0.95	0.73¢	$16.05	$52.25
95% Confidence Interval			(12.63-19.40)	(43.10-61.40)
Percent of Total Value			72.0%	74.0%
Improvement from Moderate to No Pollution (B to A)	$0.37	0.27¢	$6.19	$18.38
Percent of Total Value			28.0%	26.0%
Total Recreation Value (C to A)	$1.32	1.00¢	$22.30	$70.63
95% Confidence Interval			(18.28-25.90)	(60.44-80.82)
Number Reporting	82	86		
		Fort Collins		
Improvement from Severe to Moderate Pollution (C to B)	$1.33	1.13¢	$22.46	$65.63
95% Confidence Interval			(11.99-33.05)	(47.08-84.19)
Percent of Total Value			61.6%	63.0%
Improvement from Moderate to No Pollution (B to A)	$0.83	0.50¢	$14.02	$38.52
Percent of Total Value			38.4%	37.0%
Total Recreation Value (C to A)	$2.16	1.63¢	$36.48	$104.16
95% Confidence Interval			(16.62-56.48)	(71.89-136.45)
Number Reporting	78	89		

81

South Platte River Basin

Improvement from Severe to Moderate Pollution (C to B)	$1.48	1.17¢	$17.74	$55.91
Percent of Total Value			67.7%	70.1%
Grouped t-value[a]			1.32	1.17
Improvement from Moderate to No Pollution (B to A)	$0.70	0.48¢	$8.45	$23.87
Percent of Total Value			32.3%	29.9%
Total Recreation Value (C to A)	$2.18	1.65¢	$26.19	$79.78
Number Reporting	160		175	
Grouped t-value[a]			1.04	1.65

[a]Tests the statistical significance of the difference between grouped mean values. At the 5 percent level, there is no significant difference between the Denver and Fort Collins mean recreation values.

Table 5.5. Effect of Delay from 1983 to 2000 on Resident Household Willingness to Pay Additional Water Service Assessments and Sales Taxes to Improve Water Quality for Recreation in the South Platte River Basin, Colorado (mean values in 1981 dollars).

Water Quality Improvement Values	Water Fee Per Month (dollars)	Tax Rate (cents)	Annual Dollars Water Fee	Annual Dollars Sales Tax
	Denver Metropolitan Area			
Improved Water Quality by 1983	$1.86	1.41¢	$22.30	$70.63
95% Confidence Interval			(18.28-26.31)	(60.44-80.84)
Number Reporting	82	85		
Improved Water Quality by 2000	$1.54	1.25¢	$18.41	$61.05
95% Confidence Interval			(14.65-22.07)	(50.10-72.00)
Number Reporting	86	86		
Change in Willingness to Pay	$-0.32	-0.16¢	$-3.89	$-9.58
Percent Change	-17.6%	-13.6%	-17.4%	-13.6%
Paired t-value[a]			3.48	2.90
	Fort Collins			
Improved Water Quality by 1983	$3.05	2.30¢	$36.48	$104.16
95% Confidence Interval			(16.60-59.39)	(71.89-136.45)
Number Reporting	78	89		
Improved Water Quality by 2000	$2.40	$1.72¢	$28.72	$75.50
95% Confidence Interval			(14.84-42.44)	(53.11-97.90)
Number Reporting	78	89		
Change in Willingness to Pay	$-0.65	-0.58¢	$-7.76	$-28.66
Percent Change	-21.3%	-25.2%	-21.3%	-27.5%
Paired t-value[a]			2.16	2.95
	South Platte River Basin			
Improved Water Quality by 1983	$2.19	1.65¢	$26.18	$79.78
Number Reporting	160	174		
Grouped t-value[b]			1.17	1.32

Improved Water Quality by 2000	$1.76	1.37¢	$21.11	$64.99
Number Reporting	164	175		
Grouped t-value[b]			0.92	0.72
Change in Willingness to Pay	$-0.43	-0.28¢	$-5.07	$-14.79
Percent Change	-19.4%	-16.2%	-18.5%	-18.5%

[a] Tests the significance of the difference between the two reported mean recreation values. At the 5 percent level, the mean values are significantly different.

[b] Tests the statistical significance of the difference between the grouped mean values. At the 5 percent level, there is no significant difference between the Denver and Fort Collins reported mean recreation values.

for Denver and Fort Collins. As water pollution abatement is delayed, water quality values in Fort Collins fall at a rate about twice as fast as Denver. The differences in values reported in the two cities are significant at the 5 percent level.

Understanding the effects of a delay in the improvement of water quality on recreation values is important, since it appears that the goals established for water quality will not be met by 1983 in the South Platte River Basin. The Environmental Protection Agency[20] reported that:

> Current water quality in the South Platte River and its tributaries is generally poor. Relatively good water quality is found in streams' headwaters at the fringes of the urbanized Denver region, but water quality deteriorates as the streams flow through the urban area. By the time the South Platte reaches Henderson downstream of Denver, water quality closely resembles the treated discharge from a sewage treatment plant. The alternative strategies evaluated in the EIS all result in improved water quality by 1983, but the goals established for water quality are not met.

RIVER BASIN VERSUS STATE VALUES

Results of this study suggest that the river basin is an appropriate geographic area when evaluating recreation benefits from improved water quality. Residents of the river basin were asked how much their willingness to pay for improved water quality would change if waterways of the entire state were improved to level A. Table 5.6 shows that the average willingness to pay to improve water quality throughout Colorado was slight by comparison to the river basin where residents live. The paired t-test showed no significant difference at the 5 percent level. In other words, it seems likely that residents of the river basin would not be willing to pay for improved water quality in other river basins in the state. However, residents of other river basins throughout Colorado may be willing to pay to improve water quality in their own local area. The implication of these results is that benefit values for the South Platte River Basin are additive to water quality values which could be estimated for each major river basin in the state.

Residents of the South Platte River Basin tend to engage in water-based recreation activity within the basin. Denver residents reported an average of twenty water-based recreation activity days annually of which twelve or some 60 percent were within the river basin. Fort Collins residents reported an average of twenty-six water-based recreation activity days annually of which twenty-one or about 80 percent were within the river basin. Still, with 20 to 40 percent of annual water-based recreation activities outside the river basin, it is surprising that residents were unwilling to pay for improved water quality at these other locations. This may be related, in part, to the opinions respondents hold concerning who should pay for water quality improvement.

Table 5.6. South Platte River Basin (SPRB) Resident Household Willingness to Pay Additional Water Service Assessments and Sales Taxes to Improve Water Quality for Recreation Throughout Colorado (mean values in 1981 dollars).

Water Quality Improvement Values	Water Fee Per Month (dollars)	Tax Rate (cents)	Annual Dollars	
			Water Fee	Sales Tax

Denver Metropolitan Area

Improved Water Quality in the SPRB	$1.86	1.41¢	$22.30	$70.63
95% Confidence Interval			(18.28-26.31)	(60.44-80.84)
Number Reporting	82	85		
Improved Water Quality in Colorado	$1.82	1.40¢	$21.79	$68.80
95% Confidence Interval			(17.89-25.58)	(58.35-79.26)
Number Reporting	82	86		
Change in Willingness to Pay	$-0.04	-0.01¢	$-0.51	$-1.83
Percentage Change	-2.3%	-0.01%	-2.3%	-2.2%
Paired t-value[a]			0.85	0.72

Fort Collins

Improved Water Quality in the SPRB	$3.05	2.30¢	$36.48	$104.16
95% Confidence Interval			(16.60-56.39)	(71.89-136.45)
Number Reporting	78	89		
Improved Water Quality in Colorado	$3.10	2.40¢	$37.16	$107.12
95% Confidence Interval			(16.79-57.60)	(74.18-140.06)
Number Reporting	78	89		
Change in Willingness to Pay	$0.05	0.10¢	$0.68	$1.03
Percent Change	0.02%	0.04%	0.02%	2.84%
Paired t-value[a]			-0.58	-1.24

Table 5.6. (continued)

Water Quality Improvement Values	Water Fee Per Month (dollars)	Tax Rate (cents)	Annual Dollars	
			Water Fee	Sales Tax
	South Platte River Basin			
Improved Water Quality in the SPRB	$2.19	1.65¢	$26.18	$79.78
Number Reporting	160	174		
Grouped t-value[b]			1.17	1.32
Improved Water Quality in Colorado	$2.17	1.66¢	$26.01	$79.28
Number Reporting	160	175		
Grouped t-value[b]				
Change in Willingness to Pay	$-0.02	0.01¢	$-0.17	1.89 $-0.50
Percent Change	-0.65%	0.85%	-0.65%	-0.64%

[a]Tests the statistical significance of the difference between the two reported mean recreation values. At the 5 percent level, there is no significant difference between the values.
[b]Tests the statistical significance of the difference between grouped mean values. At the 5 percent level, there is no significant difference between the Denver and Fort Collins reported mean recreation values.

WHO SHOULD BEAR THE COST OF IMPROVED WATER QUALITY?

When asked who should bear the costs of water quality, the modal response in both Denver and Fort Collins was the community (city residents) as a whole. As shown in Table 5.7 this response was more frequent in Denver (42.5 percent) than in Fort Collins (31.7 percent). Both cities derive a large part of their employment from industrial and manufacturing firms although the impact of these firms is greater in Denver. One explanation for this result may be provided by the case of an electronics plant which is located in a town near Fort Collins. Although it is a source of heavy metal discharge into the Big Thompson River, many residents may feel that cleanup costs should be spread over all segments of the population so as to minimize the effects of job losses from pollution abatement. For this part of the population, the benefits of employment probably outweigh individual cost of pollution abatement when dispersed throughout the community. Others who suggested the community as a whole should be held responsible may feel that although they are benefactors of clean water or consumers of goods from polluting industries, they should not be held solely responsible for treatment costs.

The second most common response as to who should pay for water quality was some combination of the first three categories with polluting industries bearing primary responsibility. About 22 percent of the Denver residents reported this opinion compared with 17.8 percent of the Fort Collins residents. Polluting firms bear primary responsibility according to 13.9 percent of Denver residents and 16.8 percent of those interviewed in Fort Collins.

The most infrequent choices of who should pay for improved water quality were the final consumer and some combination of the first three categories with primary responsibility held by the final consumer. Apparently almost all residents are aware that they are consumers of products made by polluting industries. Still, they do not feel that the costs of water quality preservation should be chiefly borne by them. Although they realize at least some of the costs will eventually be passed on to them, they may feel that requiring beneficiaries and polluting firms to assist in the financing of water quality improvement may provide stronger incentives to reduce pollution than through financing primarily by passing water quality costs on to the consumer of goods produced by polluting industries. Only 1 percent of the total sample chose either of these methods of finance.

ENVIRONMENTAL AWARENESS

Table 5.8 suggests that residents of Denver and Fort Collins were quite sensitive to their environmental surroundings. In both Denver and Fort Collins, about 84 percent of the residents interviewed reported that they are aware of regional environmental problems such as air, water, and solid waste pollution. One-half of the Denver residents and 37 percent of the Fort Collins residents reported that environmental problems have affected their family or others in the area. These environmental problems included human health, unhealthy plants or animals, reduced quality of living, or lowered

Table 5.7. Resident Opinions as to Who Should Pay for Water Quality in the South Platte River Basin, Colorado, 1976.

Who Should Pay	Denver Metropolitan Area		Fort Collins		South Platte River Basin	
	Number	Percent	Number	Percent	Number	Percent
The people benefiting	7	6.9	6	5.9	13	6.4
The final consumer	0	0	2	2.0	2	1.0
The polluting firms	14	13.9	17	16.8	31	15.4
Some combination of the above	5	5.0	15	14.9	20	9.9
Some combination of the above, the people benefiting bear primary responsibility	0	0	3	3.0	3	1.5
Some combination of the above, the final consumer bears primary responsibility	2	2.0	0	0	2	1.0
Some combination of the above, the polluting firms bear primary responsibility	22	21.8	18	17.8	40	19.8
Some combination of the above, the people benefiting and the polluting firms bear primary responsibility	5	5.0	6	5.9	11	5.4
Some combination of the above, the final consumer and the polluting firms bear primary responsibility	3	2.9	2	2.0	5	2.5
The community as a whole bears primary responsibility	43	42.5	32	31.7	75	37.1

Table 5.8. Environmental Awareness of Residents in the South Platte River Basin, Colorado, 1976.

Environmental Concern	Denver Metropolitan Area	Fort Collins	South Platte River Basin
	Percent of Sample		
Awareness of Environmental Problems in the Area	84.2	83.2	83.9
Knowledge of Environmental Damages			
Personal Knowledge	50.0	36.6	46.3
Respondent's Household	11.2	5.0	9.5
Other Households	4.1	5.9	4.6
Both Respondent's Household and Other Households	34.7	25.7	32.3
No Personal Knowledge	50.0	63.4	53.7
Quality Rating of South Platte River Basin			
Poor	18.2	7.9	15.4
Fair	48.5	43.6	47.2
Good	30.3	42.6	33.7
Excellent	3.0	5.9	3.8
Importance of South Platte River Basin for Recreation Use			
Very Important	20.0	29.7	22.7
Important	35.0	40.6	36.5
Somewhat Important	29.0	14.9	25.2
Not Important	16.0	14.9	15.7
Preferred Distribution of Increased Federal Revenues			
Water Quality	21.3	23.4	21.9
Air Quality	19.5	19.6	19.5
Health Services	17.4	17.7	17.5
Education	17.7	18.5	17.9
Highways	6.4	7.9	6.8
National Defense	10.8	8.9	10.0
Other	6.1	3.7	5.4

property values.

Over one-half of the residents of the two cities rated the quality of waterways in the South Platte River Basin as either poor or fair. About 13 percent of the respondents rated the water quality as poor, while the largest proportion of respondents (46 percent) rated the quality of water in the basin as fair. Approximately one-third of the respondents rated the quality of water in the basin as good. Only 4 percent of all respondents rated the quality of water as excellent.

The table shows that residents of both cities considered the South Platte River Basin an important resource for water-based recreation activities. In Denver, 20 percent of households report that the river basin was very important as a recreation resource compared with 30 percent in Fort Collins. The largest number of respondents reported the river basin was an important recreation area for them. Almost one-third of the households in Denver and 41 percent of households in Fort Collins reported the area was important as a source for water-based recreation activities. Only about 15 percent of the residents reported the river basin was unimportant to them as a source for water-based recreation activities.

Residents of Fort Collins rated the importance of the river basin as a water-based recreation area more highly than did residents of Denver. The Cache la Poudre River, a major tributary of the South Platte River, emerges from the mountains just above Fort Collins. Horsetooth Reservoir is also located in the foothills just west of the city. Both areas were heavily used for day outings by residents of Fort Collins. The incidence of use of the South Platte River as it passes through Denver was much less by comparison. Travel distance to popular recreation areas was also substantially greater.

Residents of both Denver and Fort Collins would be willing to allocate the greatest share of an increase in federal revenues appropriated by Congress for improved water quality. The source of increased federal revenue was not specified in the question. They would allocate over 22 percent of increased federal revenues to improved water quality, and nearly 20 percent to improved air quality.

INTER-CITY COMPARISONS

There was a large difference in the average benefits reported by Denver and Fort Collins resident households. Given the sizeable dispersion of household benefit estimates, the apparent difference between means was tested to see if it exceeds that which might occur randomly. Table 5.4 showed the results of the group t-test, comparing Denver and Fort Collins benefit values. The t-test provides an estimate of the significance of the difference in the reported mean values for the two cities. None of the t-value statistics were significant at the 5 percent level and only two at the 10 percent level. On the basis of these tests, we conclude that there is no strong evidence of a difference between the reported average benefits of Denver and Fort Collins residents. We do not find evidence in this study to support the hypothesis that residents of smaller cities would be willing to pay more for water-based recreation than residents of large cities.

Table 5.4 showed the average willingness to pay additional sales tax and water service fees for improved water quality in Denver and Fort Collins. Annual average household values were higher in Fort Collins than in Denver with the exception of the subsample of nonusers where existence and bequest values were slightly higher in Denver. This would suggest that as the size of the city is increased, the recreation use value of water quality in the South Platte River Basin may tend to decrease relative to preservation value. Resident households in Fort Collins were willing to pay about $104 annually in additional sales taxes for improved water quality for recreation use compared to about $70 for resident households in Denver, nearly 60 percent more. However, a nonuser subsample of Denver residents were willing to pay an average of $58 annually in additional sales taxes to preserve water quality for existence and bequest demands compared to $57 for residents of Fort Collins.

It was expected that differences between annual water quality values for recreation use would be related to differences in number of days of recreation use of the South Platte River Basin. Denver residents reported fewer days of water-based recreation use of the river basin than Fort Collins residents. Average household use was reported as 12.8 days in Denver compared to 20.7 days in Fort Collins. When annual water quality values were divided by the number of days the river basin was used, there was no appreciable difference between value of water quality for recreation use in the two cities. Average total recreation benefits including option value were equivalent to $7.54 per day in Denver and $7.36 per day in Fort Collins. However, the variable, days of water-based recreation use of the river basin, was not statistically significant in explaining willingness to pay for improved water quality, as will be shown in Chapter 6.

NOTES

1. U. S. Department of Commerce, Survey of Current Business, 1:4 (Washington, D.C.: U.S. Department of Commerce, April 1981).

2. David S. Brookshire, Larry S. Eubanks, and Alan Randall, "Valuing Wildlife Resources: An Experiment," Transactions, North American Wildlife Conference, 38 (1978), 302-10.

3. Christopher R. Low, "The Option Value for Alaskan Wilderness," Ph.D. dissertation, (Los Angeles: University of California, 1979).

4. Richard G. Walsh, Richard A. Gillman, and John B. Loomis, Wilderness Resource Economics: Recreation Use and Preservation Values, Department of Economics, (Fort Collins: Colorado State University, 1981).

5. Richard G. Walsh, Ray K. Ericson, John R. McKean, and Robert A. Young, Recreation Benefits of Water Quality, Rocky Mountain National Park, South Platte River Basin, Colorado, Technical Report No. 12, Colorado Water Resources Research Institute, (Fort Collins: Colorado State University, May 1978).

6. Sharon Oster, "Survey Results on the Benefits of Water Pollution Abatement in the Merrimack River Basin," Water Resources

Research, 13 (December 1977), 882-84.

7. Jack L. Knetsch and Robert K. Davis, "Comparison of Methods for Recreation Evaluation," in Water Research, Allen V. Kneese and Stephen C. Smith, eds., (Baltimore: Johns Hopkins University Press, 1966).

8. Nicolaas W. Bouwes and Robert Schneider, "Procedures in Estimating Benefits of Water Quality Change," American Journal of Agricultural Economics, 61:3 (August 1979), 535-39.

9. Freeman erroneously reported that an extrapolation of our results to estimate national recreation use and option benefits would be almost $20 billion annually in 1978 dollars. This was well above the acceptable range and he concluded that "it is dangerous to extrapolate in this way . . ." A correct extrapolation of our South Platte River Basin data would result in a national estimate within his acceptable range and very close to his most likely case. A. Myrick Freeman III, The Benefits of Air and Water Pollution Control, A Review and Synthesis of Recent Estimates, Report prepared for the Council on Environmental Quality, (Washington, D.C.: 1979), 154-55, 171.

10. Alan Randall, Barry Ives, and Clyde Eastman, "Bidding Games for Valuation of Aesthetic Environmental Improvement," Journal of Environmental Economics and Management, 1 (Fall 1974), 132-49.

11. William D. Schulze, Ralph C. d'Arge, and David S. Brookshire, "Valuing Environmental Commodities: Some Recent Experiments," Land Economics, 57:2 (May 1981), 151-72.

12. David S. Brookshire, Barry C. Ives, and William D. Schulze, "The Valuation of Aesthetic Preferences," Journal of Environmental Economics and Management, 3 (Fall 1976), 325-46.

13. Freeman, op cit., 135, 171.

14. Phillip Meyer, Recreational and Preservation Values Associated with the Salmon of the Fraser River, Information Report Series No. PAC/IN-74-1, (Vancouver, B.C.: Environment Canada, Fisheries and Marine Service, Southern Operations Branch, Pacific Region, 1974).

15. Fred W. Gramlich, "The Demand for Clean Water: The Case of the Charles River," National Tax Journal, 30:2 (June 1977), 183-94.

16. Walsh, Ericson, McKean, and Young, op cit., 52-57.

17. Peter Bohm, "Estimating the Demand for Public Goods: An Experiment," European Economic Review, 3 (June 1972), 111-30.

18. Richard G. Walsh, Jared P. Soper, and Anthony A. Prato, Efficiency of Wastewater Disposal in Mountain Areas, Technical Report No. 10, Environmental Resources Center, (Fort Collins: Colorado State University, January 1978).

19. Randall, et al., op cit.; Brookshire, et al., op cit.; and Schulze, et al., op cit.

20. U.S. Environmental Protection Agency, Draft, Denver Regional Environmental Impact Statement for Wastewater Facilities and Clean Water Plan, Summary, Region 8 (Denver, Colorado: Environmental Protection Agency, June 1977).

6
Socioeconomic Relationships

A total of thirty variables representing socioeconomic attributes of respondents were tested for significance in explaining willingness to pay for the recreation-associated value of improved water quality. Of these, fifteen variables were found to be statistically significant at the 5 percent level. Together, they explained from 17 to as much as 47 percent of the variation in willingness to pay for improved water quality. The results are summarized in Table 6.1.

Eight linear equations were fit to the thirty socioeconomic variables using least-squares multiple regression techniques. No consistent pattern was identifiable; only two variables, household income and sex, were significant in as many as five of the eight equations.

A discussion of the significant socioeconomic variables is presented in this section, along with tables showing simple cross tabulations of water quality values with the more important socioeconomic variables including: household income, sex, employment, permanence of residence, previous residence, reasons for moving, number of children, age, and recreation activities. All values are shown in 1976 dollars.

HOUSEHOLD INCOME

Table 6.2 shows household income and willingness to pay additional sales taxes for the recreation-associated value of improved water quality. The cross tabulation of the two variables suggests that average income may be positively related to the average value of water quality. As income increases, the value of improved water quality in the South Platte River Basin also tends to increase. The trend is not always consistent. For example, households with incomes at the mean of $15,000-$15,999 were willing to pay less for recreation use than households below the mean with incomes of $8,500-$10,999. Moreover, households with incomes of $21,000 and above (with an estimated average of $32,000) were willing to pay substantially more by nearly every measure. This suggests that general prosperity which increased real personal income of Colorado residents would increase the value of improved water quality in the South Platte River Basin.

Table 6.1. Regression Coefficients of Significant Socioeconomic Variables, Denver and Fort Collins, Colorado, 1976.

Significant Independent Variable[a] (5 percent level)	Recreation Use Value				Option Value			
	Sales Tax		Water Bill		Sales Tax		Water Bill	
	Denver	Fort Collins	Denver	Fort Collins	Denver	Fort Collins	Denver	Fort Collins
	Regression Coefficient							
X_1 Family Income		.00366			.00068	.00153		
X_{22} 1/Family Income	.00001						15,176	
X_{26} Family Income x Age	26.55	56.45						
X_7 Sex-Male			6.92				3.65	
X_{19} Employer-Government	28.44		28.01	7.18				
X_5 Education				8.44	3.19	38.04		2.71
X_9 Previous Residence (5,000-25,000)					7.70		5.02	
X_4 Age					-.66		-.15	
X_{13} Occupation-Professional/ Business Owner-Manager	-21.67				-14.07			
X_{24} 1/Years Lived in City			.34					
X_{20} Number of Children						-10.45		
X_{11} Previous Residence (100,000+)		45.62						
X_{14} Occupation-Housewife	20.91							
X_{15} Occupation-Retired	-30.03							
X_{17} Employer-Small Business	25.08							
Fraction of Explained Variation (R^2)	.4721	.2141	.3518	.1949	.2857	.1678	.2390	.3809

[a]See Appendix C for coding of variables.

Table 6.2. Household Income and Willingness to Pay Additional Sales Tax for Improved (C-A) Water Quality, Denver, Fort Collins, and South Platte River Basin, Colorado, 1976.

Water Quality Values	Under 6,000	6,000-8,499	8,500-10,999	11,000-13,499	13,500-15,999	Household Income Groups 16,000-18,499	18,500-20,999	21,000[a] or more	Total or Average
			Willingness to Pay Additional Sales Tax (Dollars per Year)						
Recreation Use Value									
Denver Metro Area	$22.77	$51.96	$51.08	$42.94	$42.85	$50.17	$46.33	$70.25	$50.18
Number Reporting ( )	(11)	(7)	(9)	(8)	(12)	(6)	(9)	(23)	(85)
Fort Collins	31.87	46.43	84.88	36.03	78.58	48.05	147.42	133.52	74.00
Number Reporting ( )	(13)	(10)	(10)	(16)	(13)	(5)	(6)	(16)	(89)
South Platte River Basin	25.29	50.43	60.44	41.02	52.75	49.58	73.93	87.78	56.68
Option Value									
Denver Metro Area	9.35	7.03	18.36	29.17	17.67	11.46	13.53	26.74	18.31
Number Reporting ( )	(12)	(8)	(9)	(8)	(12)	(6)	(9)	(24)	(88)
Fort Collins	28.50	32.80	52.80	15.20	16.58	10.00	46.50	63.52	34.05
Number Reporting ( )	(13)	(10)	(10)	(16)	(13)	(5)	(6)	(16)	(89)
South Platte River Basin	14.65	14.16	27.90	25.30	17.37	11.06	22.53	36.93	22.60
Total Value									
Denver Metro Area	32.13	58.99	69.44	72.11	60.52	61.63	59.86	96.99	68.49
Fort Collins	60.37	79.23	137.68	51.23	95.16	58.05	193.92	197.04	108.05
South Platte River Basin	39.94	64.59	88.34	66.32	70.12	60.64	96.46	124.71	79.28

[a] An average of $32,000.

Level of household income was significant at the 5 percent level in the regression analysis of variables associated with the value of improved water quality for recreation use. For example, in Fort Collins a $1,000 increase in household income was associated with a $3.66 increase in annual willingness to pay for improved water quality via a sales tax. Income levels and willingness to pay by Denver residents may be associated with the family life cycle. As age increases, the value of water quality increases per $1,000 of added family income, but the effect is slight. Table 6.3 shows that as age increases by 10 years, the marginal effect of a $1,000 increase in income is to increase the value of water quality by an additional 10 cents per household. In this calculation, the variation across individuals in the effects of willingness to pay of all variables listed in Appendix C are adjusted for by multiple regression.

Table 6.3. Marginal Effect of a Change of Income on Willingness to Pay Additional Sales Tax for Improved Water Quality, at Various Age Levels, Denver Metropolitan Area, Colorado, 1976.

Age of Respondent	Change in Willingness to Pay Added Sales Tax per Year per $1,000 of Added Family Income[a]
20	$0.20
30	0.30
40	0.40
50	0.50
60	0.60
70	0.70

[a]Since the variable defined as the cross-products of age and income were significant, the marginal effect of income on willingness to pay increased sales tax for improved water quality was computed by taking the derivative of the regression equation and substituting various ages. This gives

$$\frac{\partial y_6}{\partial x_1} = B_{26} x_4$$

where the variables are defined as in the table and B_{26} is the regression coefficient of the cross-products. Substituting the regression coefficent and various ages into the above equation will yield the marginal effect of income on willingness to pay.

The regression analysis showed a significant relationship between household income and option value of water quality, as measured by willingness to pay additional sales tax. The relationship was positive in both the Denver metropolitan area and Fort Collins. There was a negative correlation between household income and option value of water quality in Fort Collins, as measured by willingness to pay additional water bill. This is inconsistent with the findings regarding willingness to pay a sales tax for the recreation value of improved water quality. One possible explanation for the negative association may be a tendency for those with higher income levels in Fort Collins to have a vested interest in the pollution of water resources from economic development. They would be willing to pay little to postpone economic development merely for the option to choose to recreate in the South Platte River Basin. They may be quite sure at present they will not choose water quality over pollution from development in the future. A higher proportion of Fort Collins residents engage in water-based recreation activities outside the river basin than do Denver residents. Higher income households are more able to travel long distances to fish in Wyoming, northwestern Colorado, and other places where recreation water resources tend to be less polluted.

SEX OF RESPONDENT

Table 6.4 shows the sex of respondents and average willingness to pay additional sales tax for improved water quality. It can be seen from the cross tabulation that, on the average, men were willing to pay more for water quality than women. In regression analysis of socioeconomic variables associated with the value of improved water quality for recreation use, sex of the respondent had a significant effect. Sexual differences were significant in both cities, but the effect was greater in Fort Collins. The value to males of improved water quality for recreation use was more than double (2.4 times) the value reported by female respondents in Fort Collins. In Denver, values reported by males were nearly 60 percent greater than those of females.

The primary reason may be that men tend to engage in water-based recreation activities more than women, particularly fishing and to some extent boating. Both men and women swim in nearly equal proportions and more women than men go sightseeing, picnicking, and walking for pleasure, much of which occurs along lakes and streams.[1]

EMPLOYMENT

The type of work people do has a significant effect on the value of improved water quality for recreation use. Table 6.5 shows a cross tabulation of where people work and willingness to pay additional sales tax for improved water quality. The average values suggest that employees of government and small business are willing to pay more for water quality than either employees of large business and manufacturing or unemployed persons in Denver. However, in Fort Collins employees of large business and manufacturing are willing to pay more for improved water quality than small business and agriculture employees. Average skill levels of business employees may be higher in Fort Collins, which has several high technology firms in close proximity.

Government employees are willing to pay more for improved water quality in four of the six regression equations. Government employees in Denver were willing to pay $3.91 more for option value and $28.44 more for water quality improvement for enhanced recreational opportunities than employees of the private sector. Denver is a center of national and regional government. Government employees may

Table 6.4. Sex of Respondent and Willingness to Pay Additional Sales Tax for Improved (C-A) Water Quality, Denver, Fort Collins, and South Platte River Basin, Colorado, 1976.

Water Quality Values	Sex		Total or Average
	Female	Male	
	Willingness to Pay Sales Tax (Dollars per Year)		
Recreation Use Value			
Denver Metro Area	$42.81	$59.30	$50.18
Number Reporting ( )	(47)	(38)	(85)
Fort Collins	40.52	93.73	74.00
Number Reporting ( )	(33)	(56)	(89)
South Platte River Basin	42.18	68.10	56.68
Option Value			
Denver Metro Area	19.14	17.37	18.31
Number Reporting ( )	(47)	(41)	(88)
Fort Collins	11.02	47.63	34.05
Number Reporting ( )	(33)	(56)	(89)
South Platte River Basin	16.89	25.63	22.60
Total Value			
Denver Metro Area	61.95	76.67	68.49
Fort Collins	51.54	141.36	108.05
South Platte River Basin	59.07	94.33	79.28

have a greater awareness of environmental problems in the region. In many cases these employees work in areas of environmental concern and planning.

Table 6.6 shows the association between the type of work people do and average willingness to pay additional sales taxes for improved water quality. The average values suggest that, in Denver, housewives are willing to pay more than professionals, business owners and managers, those in other occupations, and the retired. In Fort Collins, however, professionals, business owners and managers are willing to pay more than those in other occupations, and housewives reported the lowest values, even lower than retired persons.

The type of work people do was significant in a regression analysis of variables associated with the value of improved water quality for recreation use, as measured by willingness to pay additional sales tax. In the Denver metropolitan area, regression results show that professionals and business owners and managers value water quality by $21.67 less than other occupations. Retired residents value water quality by $30.03 less than those who remain active in the work force. Housewives were willing to pay $20.91 more than those employed in other occupations. Since this variable entered only one of the six regressions, it is difficult to conclude that it may affect a similar general response from the basin population.

Table 6.5. Type of Employer Related to Willingness to Pay Additional Sales Tax for Improved (C-A) Water Quality, Denver, Fort Collins, and South Platte River Basin, Colorado, 1976.

Water Quality Values	Small Business and Agriculture	Large Business and Manufacturing	Employer Government	Other[a] and Unemployed	Total or Average
	Willingness to Pay Additional Sales Tax (Dollars per Year)				
Recreation Use Value					
Denver Metro Area	$60.02	$44.12	$66.23	$43.74	$50.18
Number Reporting ( )	(14)	(13)	(14)	(44)	(85)
Fort Collins	57.96	112.25	103.00	44.53	74.00
Number Reporting ( )	(23)	(10)	(28)	(28)	(89)
South Platte River Basin	59.45	62.72	76.42	43.96	56.68
Option Value					
Denver Metro Area	11.46	13.73	23.24	20.95	18.31
Number Reporting ( )	(16)	(13)	(15)	(44)	(88)
Fort Collins	23.04	60.95	44.63	22.90	34.05
Number Reporting ( )	(23)	(10)	(28)	(28)	(89)
South Platte River Basin	14.76	26.62	29.17	21.49	22.60
Total Values					
Denver Metro Area	71.48	57.85	89.47	64.69	68.49
Fort Collins	81.00	173.20	147.63	67.43	108.05
South Platte River Basin	74.12	89.34	105.59	63.45	79.28

[a]Other includes petro-chemicals and mining.

Table 6.6. Occupation and Willingness to Pay Additional Sales Tax for Improved (C-A) Water Quality, Denver, Fort Collins, and South Platte River Basin, Colorado, 1976.

| Water Quality Values | Occupation ||||| |
	Professionals, Business Owners, and Managers	Housewife	Retired	Other[a]	Total or Average
	Willingness to Pay Additional Sales Tax (Dollars per Year)					
Recreation Use Value					
Denver Metro Area	$52.44	$58.63	$29.07	$53.41	$50.18
Number Reporting ( )	(23)	(17)	(14)	(31)	(85)
Fort Collins	94.00	38.61	41.68	76.50	74.00
Number Reporting ( )	(33)	(11)	(11)	(32)	(89)
South Platte River Basin	63.95	53.08	32.65	59.71	56.68
Option Value					
Denver Metro Area	16.66	30.11	6.97	18.98	18.31
Number Reporting	(24)	(17)	(16)	(31)	(88)
Fort Collins	39.64	7.93	6.27	46.06	34.05
Number Reporting ( )	(33)	(11)	(11)	(34)	(89)
South Platte River Basin	23.03	23.97	6.78	26.37	22.60
Total Value					
Denver Metro Area	69.10	88.74	36.04	72.39	68.49
Fort Collins	133.64	46.54	47.95	122.56	108.05
South Platte River Basin	86.98	77.05	39.43	86.08	79.28

[a] Other includes skilled, foreman, salesman, keeper, office worker, unskilled, and student.

EDUCATION

Table 6.7 shows the association between years of education and average willingness to pay additional sales tax for improved water quality. The cross tabulation of the two variables shows that average education level attained is positively related to the value of water quality. As mean schooling increases, the mean value of improved water quality in the South Platte River Basin tends also to increase. While the trend is not wholly consistent and there are few in the sample with less than a high school education, it is clear that they valued water quality less than those who graduated from high school. High school graduates and college graduates valued water quality less than those with professional or graduate-level education beyond the college level.

Level of education was significant at the 5 percent level in regression analysis of the factors explaining the value of improved water quality for recreation use. In Fort Collins there was a positive correlation between level of education and the value of water quality, as measured by willingness to pay a higher water service fee. Each additional year of schooling was associated with a $2.71 increase in option value and a $8.44 increase in recreation use value. This suggests that educational attainment may be associated with more concern about environmental quality. No explanation is available as to why this variable was not significant in Denver.

FORMER RESIDENCE

Table 6.8 shows a cross tabulation between the size of city of previous residence and average willingness to pay additional sales taxes for improved water quality. The average values suggest the effects vary between the two cities. Rural immigrants to Denver value water quality more highly for recreation use than immigrants from other cities whether large or small, with the lowest values reported for immigrants from large cities of 100,000 people or more. In Fort Collins, immigrants from large cities valued water quality more highly than other immigrants. Rural immigrants to Fort Collins reported the lowest values.

Place of former residence was significant in a regression analysis of variables associated with the recreation use and option value of improved water quality. For Denver metropolitan area residents, the smaller the place of former residence, the more they tended to value water quality. These former residents of rural areas and small cities may have had easier access to recreation areas and hence more recreational use of lakes and streams than those from large cities. Having developed an appreciation for the natural environment, they may place special emphasis on preserving it for recreation use in the future. For residents of Fort Collins, regression analysis shows the larger the place of former residence, the more they value water quality. Perhaps those who are willing to pay more for improved water quality tend to migrate to smaller cities which have outdoor recreation resources nearby, while those who are less willing to pay for water quality tend to remain in large cities.

Table 6.7. Education and Willingness to Pay Additional Sales Tax for Improved (C-A) Water Quality, Denver, Fort Collins, and South Platte River Basin, Colorado, 1976.

Water Quality Values	Years of Education					Total or Average
	8	9-11	12	13-15	16	Over 16	
	Willingness to Pay Additional Sales Tax (Dollars per Year)						
Recreation Use Value							
Denver Metro Area	$26.00	$45.44	$46.88	$56.95	$47.30	$48.19	$50.18
Number Reporting ( )	(1)	(4)	(17)	(28)	(23)	(12)	(85)
Fort Collins	10.42	30.80	67.16	79.68	56.70	112.01	74.00
Number Reporting ( )	(3)	(5)	(21)	(21)	(12)	(20)	(89)
South Platte River Basin	21.68	41.38	52.50	63.15	49.90	65.87	56.68
Option Value							
Denver Metro Area	8.88	7.69	13.40	27.17	19.04	9.47	18.31
Number Reporting ( )	(2)	(4)	(18)	(28)	(23)	(13)	(88)
Fort Collins	2.83	3.10	40.67	22.93	29.92	55.13	34.05
Number Reporting ( )	(3)	(5)	(21)	(21)	(19)	(20)	(89)
South Platte River Basin	7.20	6.42	20.95	26.01	22.05	22.12	22.60
Total Value							
Denver Metro Area	34.88	53.13	60.28	84.12	66.34	57.66	68.49
Fort Collins	13.25	33.90	107.83	102.61	86.82	167.14	108.05
South Platte River Basin	28.88	47.80	73.45	89.16	71.95	87.99	79.28

Table 6.8. Size of Place of Previous Residence and Willingness to Pay for Improved (C-A) Water Quality, Denver, Fort Collins, and South Platte River Basin, Colorado, 1976.

Water Quality Values	Size of Place of Previous Residence				
	100,000+	25,000-100,000	5,000-25,000	Rural	Total or Average
	Willingness to Pay Additional Sales Tax (Dollars per Year)				
Recreation Use Value					
Denver Metro Area	$43.73	$52.20	$54.44	$63.96	$50.18
Number Reporting ( )	(41)	(23)	(12)	(7)	(85)
Fort Collins	107.49	55.51	85.25	31.72	74.00
Number Reporting ( )	(30)	(26)	(16)	(17)	(89)
South Platte River Basin	61.13	53.12	62.97	55.03	56.68
Option Value					
Denver Metro Area	14.97	15.62	32.64	7.00	18.31
Number Reporting ( )	(41)	(25)	(13)	(7)	(88)
Fort Collins	47.60	26.49	48.39	8.20	34.05
Number Reporting ( )	(30)	(26)	(16)	(17)	(89)
South Platte River Basin	23.87	18.63	37.00	7.33	22.60
Total Value					
Denver Metro Area	58.70	67.82	87.08	70.96	68.49
Fort Collins	155.59	82.00	133.64	39.92	108.05
South Platte River Basin	85.00	71.85	99.97	62.36	79.28

REASONS FOR MOVING

Table 6.9 shows a cross tabulation of reasons for moving to Colorado and willingness to pay additional sales taxes for improved water quality. The average values suggest that residents who immigrated to the river basin for environmental reasons may value water quality more highly than those who came for other reasons. This is not the case for residents of the Denver metropolitan area, where those who moved there for family reasons value water quality most highly. Those who moved there for economic and other reasons such as quality of the public services, valued water quality less. This is also the case in Fort Collins.

Fort Collins residents who moved there for environmental reasons place the highest value on water quality. This is sufficient to overcome the lower values by those who moved to Denver for environmental reasons, so that the overall South Platte River Basin estimate of the value of water quality for recreation use is highest for those who moved there for environmental reasons. This is in accord with the widespread belief that people move to Colorado because of its reputation for a quality living environment. However, reasons for moving to Colorado were not significant in regression analysis of variables associated with the value of improved water quality for recreation use or for option value.

PERMANENCE OF RESIDENCE

Table 6.10 shows how long people live in one place and willingness to pay additional sales tax for improved water quality. The cross tabulation of the two variables suggests that, on average, the longer people live in one place, the less they are willing to pay for water quality improvement for recreation use. The trend is not always consistent, but the newly arrived residents of less than five years were willing to pay over one-fourth more than long standing residents of eleven to twenty years. The same relationship was apparent in Fort Collins. Newly arrived residents were willing to pay more than twice as much as residents of eleven to twenty and twenty-one to forty years. This suggests that immigration of people into the state in recent decades may have increased the value of improved water quality for recreation use.

Permanence of residence was a significant variable in regression analysis of variables associated with the value of improved water quality for recreation use, as measured by willingness to pay additional water service fees. In the Denver metropolitan area, for example, additional years of residence in the city was associated with a decrease in willingness to pay additional water fees for water quality improvement. Permanence of residence was not a significant variable in the regression analysis of variables associated with the option value of improved water quality. Still, the average values shown in Table 6.10 suggest that newly arrived residents tend to report higher values for improved water quality than residents who have lived in the area for a longer time. This appears contrary to the usual effects of permanence of community residency. Sociologists suggest that community and area pride grow as length of residency

Table 6.9. Reason for Moving to Colorado and Willingness to Pay Additional Sales Tax for Improved (C-A) Water Quality, Denver, Fort Collins, and South Platte River Basin, Colorado, 1976.

Water Quality Values	Reason for Moving to Colorado				
	Environmental	Economic or Other[a]	Family/Native	School	Total or Average
	Willingness to Pay Additional Sales Tax (Dollars per Year)				
Recreation Use Value					
Denver Metro Area	$49.60	$49.54	$52.59	$47.00	$50.18
Number Reporting ( )	(23)	(39)	(20)	(31)	(85)
Fort Collins	110.71	61.94	65.98	64.96	74.00
Number Reporting ( )	(19)	(28)	(20)	(22)	(89)
South Platte River Basin	66.28	35.82	56.30	51.97	56.68
Option Value					
Denver Metro Area	13.76	17.65	25.71	17.08	18.31
Number Reporting ( )	(26)	(39)	(20)	(3)	(88)
Fort Collins	41.33	29.73	21.96	44.26	34.05
Number Reporting ( )	(19)	(28)	(20)	(22)	(89)
South Platte River Basin	21.28	21.00	24.67	24.61	22.60
Total Value					
Denver Metro Area	63.36	67.19	78.30	64.08	68.49
Fort Collins	152.04	91.67	87.94	109.22	108.05
South Platte River Basin	87.56	56.82	80.97	76.58	79.28

[a] Other includes quality of public services.

Table 6.10. Permanence of Residence and Willingness to Pay Additional Sales Tax for Improved (C-A) Water Quality, Denver, Fort Collins, and South Platte River Basin, Colorado, 1976.

Water Quality Values	1-5	6-10	11-20	21-40	Over 40	Total or Average
	\multicolumn{6}{c}{Willingness to Pay Additional Sales Tax (Dollars per Year)}					
Recreation Use Value						
Denver Metro Area	$55.58	$43.23	$44.32	$53.78	$43.83	$50.18
Number Reporting ( )	(19)	(13)	(12)	(32)	(9)	(85)
Fort Collins	108.32	70.85	46.40	41.55	22.88	74.00
Number Reporting ( )	(38)	(13)	(17)	(15)	(6)	(89)
South Platte River Basin	69.98	50.77	44.89	50.44	38.03	56.68
Option Value						
Denver Metro Area	23.33	18.51	19.61	18.31	6.70	18.31
Number Reporting ( )	(19)	(13)	(14)	(32)	(10)	(88)
Fort Collins	53.45	36.94	16.63	13.38	5.92	34.05
Number Reporting ( )	(38)	(13)	(17)	(15)	(6)	(89)
South Platte River Basin	31.67	23.62	18.78	16.96	6.48	22.60
Total Value						
Denver Metro Area	78.91	61.74	63.92	72.09	50.53	68.49
Fort Collins	161.77	107.79	63.03	54.93	28.80	108.05
South Platte River Basin	101.65	74.39	63.67	67.40	44.51	79.28

increases. Community social bonds are stronger and neighborhoods are better preserved. Thus, it would be reasonable to expect that long term residents would value improved water quality more highly than newer residents. This enigma can be partially explained by the finding of a correlation coefficient of 0.55 between lengths of residency and age. There is a tendency for water-based recreation activity to decrease with age, hence the comparison of averages can be misleading.

AGE OF RESPONDENT

Table 6.11 shows willingness to pay additional sales tax for improved water quality by age of respondent. The average values suggest that age of respondent may be negatively related to the value of water quality. As age increases, the value of improved water quality in the South Platte River Basin tends to decline. The trend is not always consistent. For example, youths of eighteen to twenty-nine years of age are not as willing to pay for recreation use as are the middle aged, thirty to forty-nine. However, middle age respondents are willing to pay more than older persons, those fifty to sixty-five years of age and those retired, over sixty-five years of age. This suggests that recent immigration of young adults may tend to increase the value of improved water quality in the river basin.

Age of the respondent was not significant at the 5 percent level in regression analysis of the factors associated with value of improved water quality for recreation use. However, in the Denver metropolitan area, there was a significant negative correlation between age and option value of water quality, as measured by willingness to pay a higher water bill. As age increased, willingness to pay declined. For each 10 year increase in age, willingness to pay declined by $6.60 per year. This would seem reasonable as young people expect to live longer than older people and have more at stake in preserving their option to engage in water-based recreation activities in the South Platte River Basin in the future.

SIZE OF HOUSEHOLD

Table 6.12 shows size of household and willingness to pay additional sales taxes for improved water quality. As can be seen from the cross tabulation of the two variables, the average value of water quality appears to be inversely related to family size. As the number of children in resident households declines, the value of improved water quality in the South Platte River Basin tends to increase. The trend is not always consistent, but it is clear that households with two or fewer children value improved water quality more than households with three or more children. The national trend is toward fewer children per household. This suggests that future changes in size of households may result in increased willingness to pay for water quality.

Size of household was not significant at the 5 percent level in a regression of variables associated with the value of improved water quality for recreation use. However, regression analysis showed a significant negative relationship between the size of household in

Table 6.11. Age and Willingness to Pay Additional Sales Tax for Improved (C-A) Water Quality, Denver, Fort Collins, and South Platte River Basin, Colorado, 1976.

Water Quality Values	Age (Years)				
	18-29	30-49	50-64	Over 64	Total or Average
	Willingness to Pay Additional Sales Tax (Dollars per Year)				
Recreation Use Value					
Denver Metro Area	$46.71	$53.67	$50.35	$48.98	$50.18
Number Reporting ( )	(24)	(27)	(22)	(12)	(85)
Fort Collins	66.43	104.07	59.93	35.54	74.00
Number Reporting ( )	(31)	(31)	(14)	(13)	(89)
South Platte River Basin	50.10	67.63	53.00	45.26	56.68
Option Value					
Denver Metro Area	27.40	21.09	13.32	5.59	18.31
Number Reporting ( )	(24)	(27)	(23)	(14)	(88)
Fort Collins	43.19	44.04	18.04	5.67	34.05
Number Reporting ( )	(31)	(31)	(14)	(13)	(89)
South Platte River Basin	31.71	27.45	14.63	5.61	22.60
Total Value					
Denver Metro Area	74.11	74.76	63.67	54.57	68.49
Fort Collins	109.62	148.11	77.97	41.21	108.05
South Platte River Basin	81.81	95.08	67.63	50.87	79.28

Table 6.12. Size of Household and Willingness to Pay Additional Sales Tax for Improved (C-A) Water Quality, Denver, Fort Collins, and South Platte River Basin, Colorado, 1976.

Water Quality Values	Number of Children				
	0	1	2	3	4 or more	Total or Average

	0	1	2	3	4 or more	Total or Average
	Willingness to Pay Additional Sales Tax (Dollars per Year)					
Recreation Use Value						
Denver Metro Area	$48.63	$53.65	$41.32	$62.67	$52.84	$50.18
Number Reporting ( )	(23)	(10)	(23)	(13)	(16)	(85)
Fort Collins	70.98	88.15	95.08	35.89	39.89	74.00
Number Reporting ( )	(27)	(10)	(32)	(13)	(7)	(89)
South Platte River Basin	54.73	63.21	56.21	55.25	49.25	56.68
Option Value						
Denver Metro Area	19.17	12.65	16.67	20.13	22.28	18.31
Number Reporting ( )	(23)	(10)	(24)	(14)	(17)	(88)
Fort Collins	45.00	69.55	25.99	10.12	22.39	34.05
Number Reporting ( )	(27)	(10)	(32)	(13)	(7)	(89)
South Platte River Basin	26.23	28.41	19.25	17.36	22.31	22.60
Total Value						
Denver Metro Area	67.80	66.30	57.99	82.80	75.12	68.49
Fort Collins	115.98	157.70	121.07	46.01	62.28	108.05
South Platte River Basin	80.96	91.62	75.46	72.61	71.56	79.28

Fort Collins and the option value of water quality, as measured by willingness to pay additional sales tax. As the number of children in households increased, option value decreased by $10.45 per year. This may be related to ability to pay. Smaller families have more income per person than larger families, at any given level of income. Thus, smaller families have more income available to pay additional sales taxes for improved water quality.

RECREATION USE

Table 16.3 shows the number of water-based recreation activity days annually in the South Platte River Basin reported by survey respondents and willingness to pay additional sales tax for improved water quality. Number of water-based recreation activity days was not significant in regression analysis of variables associated with recreation use and option value of improved water quality. The average values suggest that the relationship may be curvilinear. As recreation activity increases from zero to twenty-one days annually, the average value of water quality for recreation use also tends to increase. Over twenty-one days annually, water quality values fall off. This is particularly evident for Fort Collins residents. However, the tendency is not always consistent. Even in Fort Collins, the average option value of water quality increases continuously over the entire range of recreation activity days.

It seems that with the possible exception of values for recreation use in the Denver metropolitan area, those who are not currently engaging in water-based recreation activities in the river basin may be less willing to pay for improved water quality to enhance recreation enjoyment of its use both now and in the future. Also, light users who report current water-based recreation activity as one to seven days annually appear to have lower water quality values than either medium-heavy users of eight to twenty-one days or heavy users of over twenty-one days annually.

Table 6.14 shows the reported number of water-based recreation activity days in the United States by survey respondents and willingness to pay additional sales tax for improved water quality. It shows the same general relationship as Table 6.13.

The proportion of the population which participates in water-based recreation activities in the river basin is difficult to estimate. Of the 101 households sampled in Denver, 79.2 percent reported participating in water-based recreation activities in the South Platte River Basin, compared to 85.1 percent in Fort Collins. On this basis, the weighted average participation rate was estimated at 80.8 percent for residents of the river basin in 1976. One reason for the rather high participation rate was the definition of what constituted water-based recreation activities. It was broad including the usual water sports -- fishing, swimming, boating (power boat, sail boat, and canoe), and water skiing. Also included were non-contact recreation activities such as picnicking, sightseeing, pleasure drives, and hiking within sight of lakes and streams. Water quality was reported to enhance the enjoyment or aesthetic satisfaction of such recreation experiences.

Table 6.13. Survey Respondents Reported Annual Water-Based Recreation Activity Days in the South Platte River Basin and Willingness to Pay Additional Sales Tax for Improved (C-A) Water Quality, Denver, Fort Collins, and South Platte River Basin, Colorado, 1976.

Water Quality Values	Annual Water-Based Recreation in the South Platte River Basin (Days)				
	0	1-7	8-21	Over 21	Total or Average
	Willingness to Pay Additional Sales Tax (Dollars per Year)				
Recreation Use Value					
Denver Metro Area	$49.22	$45.66	$51.71	$54.96	$50.18
Number Reporting ( )	(17)	(22)	(32)	(14)	(85)
Fort Collins	17.38	53.24	101.85	74.57	74.00
Number Reporting ( )	(8)	(18)	(29)	(34)	(89)
South Platte River Basin	38.43	47.76	65.60	60.39	56.68
Option Value					
Denver Metro Area	12.31	18.29	22.58	16.04	18.31
Number Reporting ( )	(17)	(24)	(32)	(15)	(88)
Fort Collins	2.09	28.43	43.39	36.58	34.05
Number Reporting ( )	(8)	(18)	(29)	(34)	(89)
South Platte River Basin	8.99	21.10	28.34	21.73	22.60
Total Value					
Denver Metro Area	61.53	63.95	74.29	71.00	68.49
Fort Collins	19.47	81.67	145.24	111.15	108.05
South Platte River Basin	50.05	48.86	93.94	82.12	79.28

Table 6.14. Survey Respondents Reported Annual Water-Based Recreation Activity Days in the United States and Willingness to Pay Additional Sales Tax for Improved (C-A) Water Quality, Denver, Fort Collins, and South Platte River Basin, Colorado, 1976.

	Annual Water-Based Recreation in the United States (Days)				
Water Quality Values	0	1-7	8-21	Over 21	Total or Average
	Willingness to Pay Additional Sales Tax (Dollars per Year)				
Recreation Use Value					
Denver Metro Area	$51.22	$46.01	$52.56	$53.10	$50.18
Number Reporting ( )	(9)	(30)	(30)	(27)	(85)
Fort Collins	7.60	50.44	94.89	81.41	74.00
Number Reporting ( )	(5)	(21)	(29)	(31)	(89)
South Platte River Basin	35.44	47.24	64.29	60.94	56.68
Option Value					
Denver Metro Area	12.81	16.66	21.86	19.44	18.31
Number Reporting ( )	(9)	(32)	(30)	(28)	(88)
Fort Collins	0.95	24.94	39.03	43.68	34.05
Number Reporting ( )	(5)	(21)	(29)	(41)	(89)
South Platte River Basin	8.42	18.95	26.62	26.15	22.60
Total Value					
Denver Metro Area	64.03	62.67	74.42	72.54	68.49
Fort Collins	8.55	75.38	133.92	125.09	108.05
South Platte River Basin	48.88	66.19	90.91	87.09	79.28

This may appear to be a high estimate of participation in comparison to a 1972 U.S. Census survey which showed that in the Western Region, 26.7 percent of the population fished, 5.9 percent water skied, 2.2 percent canoed, 2.4 percent sailed, 17 percent used power boats, and 36.1 percent swam at a beach.[2] However, 48.9 percent picnicked, 44 percent went sightseeing, 40.9 percent when driving for pleasure, 23.5 percent went on nature walks, and 14.2 percent bicycled. Many people engaged in more than one water-based recreation activity, and it is not possible to show the proportion of the total population who engaged in water-based sports and the proportion who did not. It is known that most boaters also fish and most swimmers also picnic. The Colorado Division of Parks and Outdoor Recreation estimated that residents of the state devoted an average of 32 days to all outdoor recreation activities in 1971. The present study found that all water-based recreation activities accounted for 20.3 days in the case of Denver residents and 26 days for Fort Collins residents. Thus water-based recreation accounted for about 62 percent in Denver and 81 percent in Fort Collins of average total outdoor recreation activities by Colorado residents. The weighted average number of water-based recreation days by river basin residents was 21.6 days or a reasonable 67.5 percent of the average number of recreation activity days reported by Colorado residents in 1971.

The South Platte River Basin appears to be the most important location for water-based recreation activities by residents of both the Denver metropolitan area and outlying cities in the basin such as Fort Collins. For the Denver residents sampled, the average number of water-based recreation days reported for 1976 was 12.8 or 63 percent in the South Platte River Basin. For Fort Collins residents, the average number of water-based recreation days in the same year was 20.7 or 80 percent in the South Platte River Basin. The weighted average amount of water-based recreation in the river basin was 15 days or an estimated 46.9 percent of the average number of recreation activity days reported by Colorado residents in 1971.

NOTES

1. Robert L. Adams, Robert C. Lewis, and Bruce H. Drake, Outdoor Recreation, Appendix A, An Economic Analysis, (Washington, D.C.: Bureau of Outdoor Recreation, U.S. Department of the Interior, December 1973).

2. Adams, Lewis, and Drake, Ibid.

7
Summary and Conclusions

The purpose of this study was to develop and apply a procedure for measuring the benefits of improved water quality to both recreational users of the resource and the general population. Benefits included: (1) consumer surplus from enhanced enjoyment of water-based recreation, (2) option value of assured choice of recreation use in the future through avoidance of irreversible pollution by mineral and energy development, and (3) preservation value of the existence of a natural ecosystem and its bequest to future generations. The South Platte River Basin, located in northeastern Colorado, was selected as the study area. A random sample of 202 resident households in Denver and Fort Collins were interviewed in their homes in the summer of 1976. The water quality benefits originally estimated for 1976 were updated to 1981. Dollar values were estimated to have inflated by 40.76 percent over this 5 year period, based on changes in the GNP Implicit Price Deflator.

The economic benefits reported in this study were based on the contingent value approach recently recommended by the U.S. Water Resources Council as suitable for studies of recreation and environmental quality. The approach relies on the stated intentions of individuals to pay with variations in water quality depicted in color photographs. The iterative technique was used to assure that individuals reported maximum points of indifference between having the stated amount of income or level of water quality.

The South Platte River Basin of Colorado contains two-thirds of the state's population concentrated along the northern Front Range of the Rocky Mountains. Metropolitan Denver, with 1.4 million residents, is the major city in the region. Smaller basin communities include Fort Collins with about 60,000 residents, Boulder, Greeley, Loveland, Longmont, and Sterling. These urban areas are growing rapidly.

The river basin is a popular water-based recreation area. Approximately 40 percent of all water-based recreation activity in Colorado takes place there. However, water quality of the South Platte River and its tributaries is generally poor. Streams flowing through urban areas are generally not in compliance with state water quality laws while most mountain streams are somewhat cleaner.

Past and to some extent current mining practices have been detrimental to water quality. In several areas, waterways have been irreversibly polluted by metal mine drainage because of the prohibitive costs associated with zero discharge. As a result, recreation use of these areas has been severely restricted. Fish and wildlife populations have been limited by the toxic effects of water acidity and discoloration has reduced the aesthetic value of these areas. Since heavy metal discharge is characteristic of many types of effluent which limit water-based recreation activity and aesthetic enjoyment, it was selected as an index of overall basin water quality. The imminent expansion of coal and metal mining and the possibility of severe irreversible effects on water quality were also important considerations in the selection of heavy metal content as an index of water quality.

The random sample of 202 resident households was drawn from current telephone directories. The residents selected were mailed an introductory letter requesting an interview which was arranged at a mutually convenient time. Respondents were shown a series of three color photos depicting levels of water quality from worst to best. They were asked what they would pay for water quality improvement to enhance recreational opportunities from the worst situation to an intermediate situation (level C to level B) and for improvement from the worst to the best level of water quality (level C to level A). They were also asked to place a value on option, bequest, and existence demands under specified conditions of the highest level of water quality (level A). A simulated market situation was described and an iterative technique was employed in which respondents answered yes or no to hypothetical increases in each of two methods of payment. Respondents were asked to estimate the maximum amount by which they would increase sales taxes per year and water service charges per month to improve water quality in the South Platte River Basin. Payment amounts were increased or decreased from initial levels by one-quarter cent per dollar and by fifty cents per month, respectively, for the methods of payment.

The largest single benefit associated with water quality improvement was to enhance current water-based recreation activities. South Platte River Basin resident households reported willingness to pay an average of $80 annually in sales taxes to improve water quality in the river basin for recreation use. The 95 percent confidence interval around this average ranged from $65 to $95. This was the value reported by 81 percent of the sample households who expect to continue to use lakes and streams in the river basin for fishing, boating, swimming, and noncontact recreation activities such as picnicking and sightseeing near water with enhanced aesthetic satisfaction of such recreation experiences.

This study is believed to be the first empirical test of the concept of option value. Application of the contingent value technique was successful in meeting the primary study objective of measuring option value and other preservation values associated with improved water quality. River basin resident households reported willingness to pay an average of nearly $32 annually in sales taxes to improve water quality in the river basin for option demand. The 95 percent confidence interval around this average value ranged from

$22 to $42. This increased recreation use value by 40 percent. Option value was defined as the amount of money a household would be willing to pay annually for the option to choose to engage in water-based recreation in the future given the imminent expansion of mineral and energy development in the river basin would preclude recreation use without payment of the fee. The protection of water quality provides the option to make a future decision between two alternative uses of waterways in the river basin, either for water-based recreation or for wastewater discharge from industrial, mineral, and energy development, under conditions of sufficient knowledge about which use will be more beneficial.

Preservation benefits of water quality were defined to include both the value placed on the existence of a natural ecosystem and the value of its bequest to future generations. Existence value is an individual's willingness to pay for the knowledge that a natural environment exists as a habitat for fish, wildlife, and other ecosystems. Bequest value is the amount of money an individual would be willing to pay to assure access to natural environments by future generations.

Preservation benefits were estimated from a 19 percent subsample of nonrecreationists who reported a zero probability of future water-based recreation in the river basin. Nonuser households reported average total preservation benefits of $59 annually with a 95 percent confidence interval of $12-$107. This included bequest value averaging $24 annually with a range of $12-$36 and existence value averaging $35 annually with a range of $0-$71. Although existence values were not significantly different from zero for this small sample, we consider the average values reasonable and a larger sample would increase their significance. As a first approximation, existence and bequest values for the subsample of nonrecreationists were extended to all resident households.

The average total benefits of water quality to all households in the river basin were estimated as $149 annually with a 95 percent confidence interval of $82-$217. This included recreation benefits of $64 with a range of $52-$77, option value of $26 with a range of $18-$34, bequest value of $24 with a range of $12-$36, and existence value of $35 with a range of $0-$71. This weighted average was somewhat lower than the average total benefits of $171 annually to 81 percent of the households engaging in water-based recreation. However, it was substantially more than the average total benefits of $59 annually to 19 percent of the households who reported they do not expect to make recreation use of the waterways in the river basin.

Total annual benefits of water quality in the South Platte River Basin were estimated as $95.1 million in 1981. This included recreation use benefits of $33.2 million, option value of $18.8 million, existence value of $25.7 million, and bequest value of $17.4 million. These estimates were derived by multiplying average total benefits per household by the estimated 730,000 households living in the river basin in 1981.

The present value of a continuous stream of annual benefits from water quality in the South Platte River Basin was calculated as $1.3 billion. This included recreation use benefits of $401 million, option value of $254 million, existence value of $348 million, and bequest value of $236 million. These estimates were derived by dividing total annual benefits in 1981 by the Federal discount rate of 7 3/8 percent used in calculation of benefits and costs of public water resource projects in 1981. This is likely to prove a conservative estimate because of growth in population and income, and the exclusion of tourist benefits, although the latter account for 30-40 percent of total water-based recreation in the river basin.

Households in the river basin indicated a greater willingness to pay for improved water quality when the method of hypothetical payment was an increase in sales taxes rather than an increase in water service fees. Willingness to pay water service fees was about one-third as much as willingness to pay additional sales taxes. We hypothesize that households were more reluctant to participate in the water service fee estimation procedure because they perceived inequities involved in its payment. Everyone, including tourists, pays sales taxes whereas only property owners and indirectly renters pay water service bills. This is the free rider problem in which tourists tend to escape payment. Thus, responses based on sales taxes were considered more equitable and a more acceptable measure of benefits. Willingness to pay water service fees were reported in this study to illustrate the fact that selection among alternative methods of hypothetical payment affects the resulting values obtained in the contingent value approach.

Improving water quality from polluted level C to intermediate quality level B accounted for $56 or 70 percent of recreation benefits estimated as $80 per household annually for improving water quality from polluted level C to clean water level A. Households interviewed were not asked to estimate the effect of intermediate water quality level B on option, existence, and bequest values. Presumably they would exhibit a similar pattern. Still, with only three data points, it would be heroic to generalize about the nature of the slope of the benefits curve for water quality. The average benefits reported for recreation use suggest that the benefits function for improving water quality would increase at a decreasing rate which is consistent with the decreasing marginal utility of consumption observed for private consumption goods.

The estimates of willingness to pay for improved water quality throughout this report were based on the premise that all waterways in the river basin would be cleaned up by 1983 and then maintained in a clean state indefinitely. If circumstances such as postponement of environmental quality objectives resulted in delaying the improvement of water quality in the river basin to the year 2000, the proportion of respondents willing to pay some amount of additional sales tax for improved water quality to enhance recreation enjoyment would decline by 11 percent. If it is not possible to improve water quality in the river basin until the year 2000, annual willingness to pay for recreation use would fall by an average of $16 annually per household or 17.4 percent from the original $80 to $64.

Results of this study suggest that the river basin is an appropriate geographic area when evaluating willingness to pay for improved water quality. When respondents were asked what they would pay for water quality improvement throughout Colorado, including the

South Platte River Basin, for enhanced recreation use, the reported values were not significantly different from South Platte River Basin values. It appears that residents of the river basin were not willing to pay directly for improved water quality in other river basins in the state.

Nearly 40 percent of the residents of the river basin were of the opinion that the community as a whole should bear the primary responsibility for paying the costs of water quality improvement. An additional 15 percent reported the opinion that the polluting industries should pay the costs, while 30 percent favored sharing the costs between polluting industries and the people benefiting. Residents of other river basins in the state may be willing to pay to improve water quality in their local areas. River basin benefits appear to be additive throughout the state and nation.

The hypothesis that size of city may affect willingness to pay for improved water quality was not supported by this study. There was no significant difference in the recreation use value and option value reported by the two cities at the 95 percent confidence level. Average recreation and option benefits were higher in Fort Collins than in Denver while existence and bequest benefits were slightly higher in Denver. However, when recreation and option benefits were divided by the number of days of water-based recreation by households in the two cities, there was no appreciable difference between the two cities. Average total recreation benefits including option value were equivalent to $7.54 per day for Denver households compared to $7.36 per day for Fort Collins households.

The expectation that the amount of recreation activity would be associated with willingness to pay for improved water quality was not supported by this study. The relationship between the number of water-based recreation activity days annually in the South Platte River Basin and willingness to pay for improved water quality was not significant at the 95 percent confidence level. Average values suggested that the relationship may have been curvilinear. As recreation activity increased from zero to twenty-one days annually, average water quality values also tended to increase. Over twenty-one days annually, water quality values fell off. However, the tendency was not always consistent. The average option value of water quality increased continuously over the entire range of recreation use, but changes in value were not statistically significant.

Income was positively related to willingness to pay for improved water quality. Level of household income was significant at the 5 percent level in the regression analysis of variables associated with the value of improved water quality for recreation use. For example, in Fort Collins a $1,000 increase in household income was associated with a $3.66 increase in willingness to pay sales tax for improved water quality (in 1976 dollars). Regression analysis showed a significant positive relationship between household income and option value of water quality, in both cities, as measured by willingness to pay additional sales tax. However, in Fort Collins, there was a negative correlation between household income and option value of water quality, as measured by willingness to pay additional water service fees. A higher proportion of Fort Collins residents reported water-based recreation activities outside of the river basin than was reported by Denver residents.

Where people work and the type of work they do had a significant effect on the value of improved water quality for recreation. Employees of small business and government were willing to pay more for water quality than either employees of large business and manufacturing or unemployed persons. The lowest values were among the retired. Employees of small business in Denver were willing to pay $25.08 more sales tax for improved water quality (in 1976 dollars). Government employees were willing to pay $28.44 more than respondents working in the private sector. In Denver, professionals, business owners, and managers valued water quality for recreation use by $21.67 per year less than other occupations. Retired residents valued water quality for recreation use by $30.03 less than those who remained active in the work force.

Whether the respondent was male or female had a significant effect on willingness to pay to improve water quality for recreation use. For example, men were willing to pay $26.55 more for water quality than women in Denver (in 1976 dollars). The primary reason may have been that men engage in water-based recreation activities more than women, particularly fishing and to some extent boating. Apparently women who worked outside the home were particularly reluctant to allocate more of their income to sales taxes for improved water quality because housewives who remained in the home were willing to pay $20.91 annually more than those employed in other occupations, whether male or female.

Number of children in the household was not significant at the 5 percent level in the regression of variables associated with the value of improved water quality for recreation use. However, there was a significant relationship between number of children per household in Fort Collins and the option value of water quality. As the number of children in Fort Collins households increased, option value decreased by $10.45 per year (in 1976 dollars).

Education level may be associated with more concern about the future of water quality than with current recreation use. There was a positive correlation between level of education and option value of water quality in Fort Collins. The relationship was significant at the 5 percent level. However, education was not significant at the 5 percent level in the regression analysis of factors explaining the value of improved water quality for recreation use.

Increasing age may be associated with decreasing concern about the future of water quality. Older people were less concerned with preserving their option to engage in water-based recreation activities in the river basin in the future. For example, willingness to pay a water bill declined by $6.60 with each 10 year increase in age of Denver respondents (in 1976 dollars). However, age of respondent was not significant at the 5 percent level in regression analysis of factors associated with the value of improved water quality for current recreation use.

Permanence of residence was a significant variable in regression analysis of variables associated with the value of improved water quality for recreation use. For example, Denver residents were willing to pay $3.40 less water service fees for each 10 years they lived in the city (in 1976 dollars). This appears contrary to the expected effects of permanence of residents on community pride, preservation of neighborhoods, and a quality environment. Recently arrived residents were willing to pay more than residents of eleven to twenty years and twenty-one to forty years. Thus, the immigration of young adults into the state in the past decade may have increased the value of improved water quality.

Reasons given for moving to the river basin were not significant in regression analysis. However, average values suggest that residents who immigrated to the river basin for environmental reasons may value water quality more highly than those who came for other reasons. This would be consistent with the widespread belief that many people move to Colorado because of its reputation for a quality living environment.

Size of former residence was significant in regression analysis of variables associated with the recreation use and option value of improved water quality. For Denver residents, the smaller the place of former residence, the more they tended to value water quality. Having developed an appreciation for the natural environment in rural areas, they may place special emphasis on preserving it for recreation use in the future. For residents of Fort Collins, regression analysis showed that the larger the place of former residence, the more they valued water quality. Perhaps those who are willing to pay more for improved water quality tend to migrate to smaller cities which have recreation resources nearby, while those who are less willing to pay for water quality remain in or move to large cities.

The contingent value approach was successful in meeting the objective of valuing the benefits of improved water quality. Contingent value techniques have been successfully used as a research tool for valuation of air quality in the past. The technique appears appropriate for valuation of a number of nonmarket goods, including water quality. It should be remembered, however, that the approach measures the hypothetical responses of individuals faced with hypothetical situations. Thus, considerable care must be exercised in the design of questions and the conduct of surveys to ensure the results obtained are as realistic as possible.

It should be emphasized that this study provides only a partial estimate of the benefits accruing to society from water quality improvement. Recreation and preservation benefits of tourists and others living outside of the basin were not considered. Benefits also would be generated from improved health of the population, less expensive municipal and industrial treatment, and from improved water for irrigation. The partial nature of this benefits study makes it difficult to determine the economically efficient level of water quality improvement.

Bibliography

Abel, Fred H., Dennis P. Tihansky, and Richard G. Walsh, <u>National Benefits of Water Pollution Control</u>, Preliminary Draft, (Washington, D.C.: Environmental Protection Agency, 1975).

Adams, Robert L., Robert C. Lewis, and Bruce H. Drake, <u>Outdoor Recreation, Appendix A, An Economic Analysis</u>, (Washington, D.C.: Bureau of Outdoor Recreation, U.S. Department of the Interior, December 1973).

Arosteguy, Daniel J., "Socio-Economic Based Projection of Wildlife Recreation in Colorado to 1985," Ph.D. dissertation, (Fort Collins: Colorado State University, May 1974).

Arrow, K. J. and R. C. Lind, "Uncertainty and the Evaluation of Public Investment Decisions," <u>Quarterly Journal of Economics</u>, 60 (June 1970), 364-78.

Arrow, K. J. and A. C. Fisher, "Environmental Preservation, Uncertainty, and Irreversibility," <u>Quarterly Journal of Economics</u>, 88 (May 1974), 312-19.

Battelle Memorial Institute, <u>Assessment of the Economic and Social Implications of Water Quality Improvements on Public Swimming</u>, (Columbus, Ohio: 1975).

Bell, Frederick W. and E. Ray Canterberry, <u>An Assessment of the Economic Benefits Which Will Accrue to Commercial and Recreational Fisheries from Incremental Improvements in the Quality of Coastal Waters</u>, (Tallahassee: Florida State University, 1975).

Binkley, Clark W. and W. Michael Hanemann, <u>The Recreation Benefits of Water Quality Improvement: Analysis of Day Trips in an Urban Setting</u>, EPA-600/5-78-010, Socioeconomic Environmental Studies Series, (Washington, D.C.: Environmental Protection Agency, June 1978).

Bishop, A. Bruce, "Impact of Energy Development on Colorado River Water Quality," <u>Natural Resources Journal</u>, XVII (October 1977), 649-71.

Bishop, Richard C. and Thomas A. Heberlein, "Measuring Values of Extramarket Goods: Are Indirect Measures Biased?" <u>American Journal of Agricultural Economics</u>, LXI (December 1979), 926-30.

Bohm, Peter, "An Approach to the Problem of Estimating Demand for Public Goods," <u>Swedish Journal of Economics</u>, 73 (1971), 55-66.

Bohm, Peter, "Estimating the Demand for Public Goods: An Experiment,"

European Economic Review, 3 (June 1972), 111-30.
Bohm, Peter, "Option Demand and Consumer's Surplus: Comment," American Economic Review, 65 (September 1975), 733-36.
Bohm, Peter, "Estimating Willingness to Pay: Why and How?" The Scandinavian Journal of Economics, LXXXI (1979), 142-53.
Bouwes, Nicolaas W. and Robert Schneider, "Procedures in Estimating Benefits of Water Quality Change," American Journal of Agricultural Economics, 61 (August 1979), 535-39.
Bradford, David F., "Benefit-Cost Analysis and Demand Curves for Public Goods," Kyklos, 23 (1970), 775-91.
Brookshire, David S., Barry C. Ives, and William D. Schulze, "The Valuation of Aesthetic Preferences," Journal of Environmental Economics and Management, 3 (Fall 1976), 325-46.
Brookshire, David S., Larry S. Eubanks, and Alan Randall, "Valuing Wildlife Resources: An Experiment," Transactions, North American Wildlife Conference, 38 (1978), 302-10.
Brookshire, David S. and Thomas Crocker, "The Use of Survey Instruments in Determining the Economic Value of Environmental Goods: An Assessment," in Assessing Amenity Resource Values, Rocky Mountain Forest and Range Experiment Station Report No. 68, (Fort Collins, Colo.: U.S. Forest Service, 1979).
Brown, W. G. and J. M. Hammack, Waterfowl and Wetlands, Towards Bioeconomic Analysis, (Baltimore: Johns Hopkins University Press, 1974).
Burns, M. E., "A Note on the Concept and Measure of Consumer Surplus," American Economic Review, 63 (June 1973), 335-44.
Byerlee, D. R., "Option Demand and Consumer Surplus: Comment," Quarterly Journal of Economics, 85 (August 1971), 523-27.
Cesario, Frank J., "Congestion and Valuation of Recreation Benefits," Land Economics, 56 (August 1980), 329-38.
Cicchetti, Charles J., A Primer for Environmental Preservation: The Economics of Wild Rivers and Other Natural Wonders, Module 20, (New York: MSS Modular Publications, Inc., 1974).
Cicchetti, Charles J. and V. K. Smith, "Congestion, Quality Deterioration, and Optimal Use: Wilderness Recreation in the Spanish Peaks Primitive Area," Social Science Research, 2 (March 1973), 15-30.
Cicchetti, Charles J. and A. M. Freeman III, "Option Demand and Consumer Surplus: Further Comment," Quarterly Journal of Economics, 85 (August 1971), 529-39.
Clawson, Marion, Methods of Measuring the Demand for and Value of Outdoor Recreation, Reprint No. 10, (Washington, D.C.: Resources for the Future, Inc., February 1959).
Clawson, Marion and Jack L. Knetsch, Economics of Outdoor Recreation, (Baltimore: Johns Hopkins University Press, 1966).
Colorado Department of Health, Colorado Water Quality Report: 1975, (Denver: Colorado Department of Health, April 15, 1975).
Colorado Department of Health, Demographic Profile: Colorado Planning and Management, Districts 2 and 3, (Denver: Colorado Department of Health, May 7, 1976).
Colorado Division of Mines, A Summary of Mineral Industry Activities in Colorado, 1975, (Denver: State of Colorado, 1975).
Colorado Division of Parks and Outdoor Recreation, Interim Colorado

Comprehensive Outdoor Recreation Plan, (Denver: State of Colorado, 1974).

Colorado Division of Planning, Ethnic Group Population of Colorado Counties, 1960-1976, (Denver: State of Colorado, April 23, 1976).

Conrad, Jon M., "Quasi-Option Value and the Expected Value of Information," Quarterly Journal of Economics, XCIV (1980), 813-20.

Davidson, Paul F., Gerard Adams, and Joseph Seneca, "The Social Value of Water Recreational Facilities Resulting from an Improvement in Water Quality: The Delaware Estuary," in Water Research edited by Allen V. Kneese and Stephen C. Smith, (Baltimore: Johns Hopkins University Press, 1966), 175-211.

Ditton, Robert and Thomas Goodale, Marine Recreational Use of Green Bay: A Survey of Human Behavior and Attitude Patterns, Technical Report No. 17, Sea Grant Program, (Madison: University of Wisconsin, 1972).

Dwyer, John F., John R. Kelly, and Michael D. Bowes, Improved Procedures for Valuation of the Contribution of Recreation to National Economic Development, Report No. 128, Water Resources Center, (Urbana: University of Illinois, September 1977).

Eastman, Clyde, Peggy Hoffer, and Alan Randall, A Socioeconomic Analysis of Environmental Concerns: Case of the Four Corners Electric Power Complex, Bulletin 626, Agricultural Experiment Station, (Las Cruces: New Mexico State University, September 1974).

Economic Research Service in cooperation with Missouri Agricultural Experiment Station, An Econometric Model for Predicting Water-Oriented Outdoor Recreation Demand, (Washington, D.C.: U.S. Department of Agriculture, March 1969).

Engineering Consultants, Inc. and Toups Corporation, Comprehensive Water Quality Management Plan: South Platte River Basin, Colorado, (Denver: Colorado Department of Health, November 1974).

Ericson, Raymond P., "Water Quality Values in Outdoor Recreation," Ph.D. dissertation, (Fort Collins: Colorado State University, 1977).

Fischer, David W., "Willingness to Pay as a Behavioral Criterion for Environmental Decision-Making," Journal of Environmental Management, 3 (1975), 29-41.

Fisher, A. C. and J. V. Krutilla, "Valuing Long-Run Ecological Consequences and Irreversibilities," Journal of Environmental Economics and Management, 1 (August 1974), 96-108.

Fisher, A. C., J. F. Krutilla, and C. J. Cicchetti, "The Economics of Environmental Preservation: A Theoretical and Empirical Analysis," American Economic Review, 62 (September 1972), 605-19.

Fisher, A. C. and Frederick M. Peterson, "The Environment in Economics: A Survey," Journal of Economic Literature, 14 (March 1976), 1-33.

Freeman, A. Myrick III, "Benefits of Pollution Control," in Critical Review of Estimating Benefits of Air and Water Pollution Control edited by A. Hershaft, Report to the Environmental Protection Agency, (Rockville: Enviro Control, Inc., October 1976), 11.1-11.55.

Freeman, A. Myrick III, The Benefits of Air and Water Pollution Control: A Review and Synthesis of Recent Estimates, Report prepared for the Council on Environmental Quality, (Washington, D.C.: 1979).

Freeman, A. Myrick III, The Benefits of Environmental Improvement: Theory and Practice, (Baltimore: Johns Hopkins University Press, 1979).

Friedman, Milton, Capitalism and Freedom, (Chicago: University of Chicago Press, 1962).

Friedman, Milton and L. J. Savage, "The Utility Analysis of Choices Involving Risk," Journal of Political Economy, 56 (August 1948), 279-304.

Gloyna, Earnest F., "Major Research Problems in Water Quality," in Water Research edited by Allan V. Kneese and Stephen C. Smith, (Baltimore: Johns Hopkins University Press, 1966).

Gordon, Irene M. and Jack Knetsch, "Consumer's Surplus Measures and the Evaluation of Resources," Land Economics, LV (February 1979), 1-10.

Gramlich, Fred W., "The Demand for Clean Water: The Case of the Charles River," National Tax Journal, 30 (June 1977), 183-94.

Greenley, Douglas A., "Recreation and Preservation Benefits from Water Quality Improvement," Ph.D. dissertation, (Fort Collins: Colorado State University, 1979).

Greenley, Douglas A., Richard G. Walsh, and Robert A. Young, "Option Value: Empirical Estimates from a Case Study of Recreation and Water Quality," Quarterly Journal of Economics, (November 1981).

Grubb, Herbert W. and James T. Goodwin, Economic Evaluation of Water-Oriented Recreation, Report No. 84, Preliminary Texas Water Plan, (Austin: Texas Water Development Board, 1968).

Heintz, H. T., A. Hershaft, and G. C. Horak, National Damages of Air and Water Pollution, Report to the Environmental Protection Agency, (Rockville: Enviro Control, Inc., 1976).

Henderson, James M. and Richard G. Quandt, Microeconomic Theory: A Mathematical Approach, (New York: McGraw-Hill, Inc., 1971).

Henry, Claude, "Investment Decisions Under Uncertainty: The 'Irreversibility Effect'," American Economic Review, 64 (December 1974), 1006-12.

Henry, Claude, "Option Values in the Economics of Irreplaceable Assets," The Review of Economic Studies: Symposium on the Economics of Exhaustible Resources, (1974), 89-104.

Hotelling, Harold, "The Economics of Public Recreation," The Prewitt Report, Land and Recreation Planning Division, National Park Service, (Washington, D.C.: U.S. Department of the Interior, 1949).

Kalter, R. J. and L. E. Gosse, Outdoor Recreation in New York State: Projection of Demand, Economic Value, and Pricing Effects for the Period 1970-85, Special Cornell Service No. 5, (Ithaca, N.Y.: Cornell University, 1969).

Kneese, A. V. and B. T. Bower, eds., Environmental Quality Analysis: Theory and Method in the Social Sciences, (Baltimore: Johns Hopkins University Press, 1972).

Kneese, A. V. and C. W. Schultze, Pollution, Prices, and Public Policy, (Washington, D.C.: Brookings Institution, 1975).

Knetsch, Jack L., "Displaced Facilities and Benefit Calculations," Land Economics, 53 (February 1977), 123-29.

Knetsch, Jack L., and Robert K. Davis, "Comparison of Methods for Recreation Evaluation," in Water Research edited by Allan V. Kneese and Stephen C. Smith, (Baltimore: Johns Hopkins University Press, 1966).

Krutilla, John V., "Conservation Reconsidered," American Economic Review, 57 (September 1967), 777-86.

Krutilla, John V. and Anthony C. Fisher, The Economics of Natural Environments, (Baltimore: Johns Hopkins University Press, 1975).

Krutilla, John V., C. J. Cicchetti, A. M. Freeman, III, and C.S. Russell, "Observations on the Economics of Irreplaceable Assets," in Environmental Quality Analysis: Theory and Method in Social Science edited by A. V. Kneese and B. T. Bower, (Baltimore: Johns Hopkins University Press, 1972), 69-112.

Krutilla, John V. and C. J. Cicchetti, "Evaluating Benefits of Environmental Resources with Special Application to Hell's Canyon," Natural Resources Journal, 12 (January 1972), 11-29.

Lerner, Lionel J., "Quantitative Indices of Recreational Values," Water Resources and Economic Development of the West: Economics in Outdoor Recreation Policy, Report No. 11, Conference Proceedings of the Western Agricultural Economics Research Council, jointly with the Western Farm Economic Association, (Reno: University of Nevada, 1962).

Leuzzi, L. and R. Pollock, "Option Demand: A Mixed Goods Case," Public Finance, 31 (1976), 396-405.

Lindsay, C. M., "Option Demand and Consumer's Surplus," Quarterly Journal of Economics, 83 (May 1969), 344-45.

Liu, Ben-chieh, "Recreation Benefit Estimation for Lake Water Quality Improvement: A Comparative Analysis," Annual Conference of the American Water Resource Association, (Minneapolis, Minn.: October 1980).

Long, M. F., "Collective Consumption Services of Individual Goods: Comment," Quarterly Journal of Economics, 81 (May 1967), 351-52.

Low, Christopher, "The Option Value for Alaskan Wilderness," Ph.D. dissertation, (Los Angeles: University of California, 1979).

Lynch, Thomas, "Population, Pollution, Limit Fishing," Coloradoan, (Fort Collins, Colo.: 1970).

Mathews, B. S. and Gardiner S. Brown, Economic Evaluation of the 1967 Salmon Fisheries of Washington, Technical Report No. 2 (Olympia: Washington Department of Fisheries, 1970).

McKean, John R., "On Revising Micro Foundations for Analyzing Socioeconomic Organizations," The Journal of Behavioral Economics, 5 (1976), 177-85.

Meyer, Phillip A., A Comparison of Direct Questioning Methods for Obtaining Dollar Values for Public Recreation and Preservation, Technical Report Series PAC/T-75-6, (Vancouver, B.C.: Environment Canada, Fisheries and Marine Service, Southern Operations Branch, Pacific Region, 1975).

Meyer, Phillip A., Recreational and Preservation Values Associated with the Salmon of the Fraser River, Information Report Series No. PAC/IN-74-1, (Vancouver, B.C.: Environment Canada,

Fisheries and Marine Service, Southern Operations Branch, Pacific Region, 1974).

Mills, Edwin and Daniel L. Feenberg, Measuring the Benefits of Water Pollution, Report to the Environmental Protection Agency, (Washington, D.C.: 1979).

Morgan, Robert E. and Dennis A. Wentz, Effects of Metal-Mine Drainage on Water Quality in Selected Areas of Colorado, 1972-73, Colorado Water Resources Circular No. 21, (Denver: Colorado Water Conservation Board, 1974).

National Commission on Water Quality, Staff Report, (Washington, D.C.: 1976).

National Planning Association, Water-Related Recreation Benefits Resulting from P.L. 92-500, (Washington, D.C.: 1975).

Nemerow, Nelson L. and Robert Faro, Benefits of Water Quality Enhancement, (Washington, D.C.: Environmental Protection Agency, December 1970).

Oster, Sharon, "Survey Results on the Benefits of Water Pollution Abatement in the Merrimack River Basin," Water Resources Research, 13 (December 1977), 882-84.

Peskin, H. M. and E. P. Seskin, Cost Benefit Analysis and Water Pollution Policy, (Washington, D.C.: The Urban Institute, 1975).

Randall, Alan, Letter to Professor Richard G. Walsh, Department of Economics, Colorado State University, Fort Collins, May 13, 1975.

Randall, Alan, "Quantifying the Unquantifiable: Benefits from Abatement of Aesthetic Environmental Damage," Paper presented at the annual conference of the American Agricultural Economics Association, College Station, Texas, August 1974.

Randall, Alan, Barry Ives, and Clyde Eastman, "Bidding Games for Valuation of Aesthetic Environmental Improvement," Journal of Environmental Economics and Management, 1 (Fall 1974), 132-49.

Reiling, S. D., K. C. Gibbs, and H. H. Stoevener, Economic Benefits from an Improvement in Water Quality, (Washington, D.C.: Environmental Protection Agency, January 1973).

Russell, Clifford S., "Municipal Evaluation of Regional Water Quality Management Proposals," in Models for Managing Regional Water Quality edited by Robert Dorfman, Henry D. Jacoby, and Harold A. Thomas, Jr., (Cambridge, Mass.: University Press, 1972).

Samuelson, P. A., "The Pure Theory of Public Expenditure," Review of Economics and Statistics, 36 (November 1954), 387-89.

Schmalensee, R., "Option Demand and Consumer's Surplus: Valuing Price Changes Under Uncertainty," American Economic Review, 62 (December 1972), 814-24.

Schmalensee, R., "Option Demand and Consumer's Surplus: Reply," American Economic Review, 65 (September 1975), 737-39.

Schulze, William D., Ralph C. d'Arge, and David S. Brookshire, "Valuing Environmental Commodities: Some Recent Experiments," Land Economics, 57 (May 1981), 151-72.

Seckler, David W., "Analytical Issues in Demand Analysis for Outdoor Recreation: Comment," American Journal of Agricultural Economics, 50 (February 1968), 147-51.

Seckler, David W., "On the Uses and Abuses of Economic Science in the Evaluation of Outdoor Recreation," Land Economics, 42 (November

1966), 485-94.

Sinden, J. A., "The Evaluation of Extra-Market Benefits: A Critical Review," World Agricultural Economics and Rural Sociology Abstracts, 9 (1967), 1-16.

Sinden, J. A., "A Utility Approach to the Valuation of Recreational and Aesthetic Experiences," American Journal of Agricultural Economics, 56 (February 1974), 61-72.

Stevens, Joe B., "Recreation Benefits from Water Pollution Control," Water Resources Research, 2 (Second Quarter 1966), 167-82.

Stoevener, Herbert H. and William G. Brown, "Analytical Issues in Demand Analysis for Outdoor Recreation," Journal of Farm Economics, 49 (December 1967), 1295-1304.

Stoevener, Herbert H., "Analytical Issues in Demand Analysis for Outdoor Recreation: Reply," American Journal of Agricultural Economics, 50 (February 1968), 151-53.

Stoevener, H. H., J. B. Stevens, H. J. Horton, A. Sokoloski, L. P. Parrish, and E. N. Castle, Multi-Disciplinary Study of Water Quality Relationships: A Case Study of Yaquina Bay, Oregon, Special Report 348 (Corvallis: Oregon State University, February 1972).

Sutherland, Ronald J., A Regional Recreation Demand and Benefits Model, Draft Report, Environmental Research Laboratory, (Corvallis, Ore.: Environmental Protection Agency, September 1981).

Tihansky, Dennis P., "Recreational Welfare Losses from Water Pollution Along U.S. Coasts," Journal of Environmental Quality, 3 (October-December 1974), 335-46.

Unger, Samuel F., National Benefits of Achieving the 1977, 1983, and 1985 Water Quality Goals, (Manhattan, Kan.: Development Planning and Research Associates, Inc., April 1976).

University of Idaho, Sport Fishery Economics, Report to National Marine Fisheries Services, (Moscow: University of Idaho, March 1973).

U.S. Department of Commerce, Survey of Current Business, 1(4), (Washington, D.C.: U.S. Department of Commerce, April 1981).

U.S. Environmental Protection Agency, Draft, Denver Regional Environmental Impact Statement for Wastewater Facilities and Clean Water Plan, Summary, Region 8, (Denver, Colo.: Environmental Protection Agency, June 1977).

U.S. Internal Revenue Service, 1975 U.S. Individual Income Tax Return, (Washington, D.C.: Government Printing Office, 1974).

U.S. Water Resources Council, Procedures for Evaluation of National Economic Development (NED) Benefits and Costs in Water Resource Planning, (Washington, D.C.: Federal Register, December 14, 1979).

Walsh, Richard G., "Recreational User Benefits from Water Quality Improvement," Outdoor Recreation: Advances in Application of Economics, General Technical Report WO-2, (Washington, D.C.: Forest Service, U.S. Department of Agriculture, March 1977).

Walsh, Richard G., D. A. Greenley, R. A. Young, J. R. McKean, and A. Prato, Option Values, Preservation Values, and Recreational Benefits of Improved Water Quality: A Case Study of the South Platte River Basin, Colorado, Socioeconomic Environmental

Studies Series, EPA-600/5-78-001, (Research Triangle Park: Environmental Protection Agency, January 1978).

Walsh, Richard G., Jared P. Soper, and Anthony A. Prato, <u>Efficiency of Wastewater Disposal in Mountain Areas</u>, Environmental Resources Center Technical Report No. 10, (Fort Collins: Colorado State University, January 1978).

Walsh, Richard G., R. K. Ericson, J. R. McKean, and R. A. Young, <u>Recreation Benefits of Water Quality, Rocky Mountain National Park, South Platte River Basin, Colorado</u>, Technical Report No. 12, Colorado Water Resources Research Institute (Fort Collins: Colorado State University, May 1978).

Walsh, Richard G., Richard A. Gillman, and John B. Loomis, <u>Wilderness Resource Economics: Recreation Use and Preservation Values</u>, Report to the American Wilderness Alliance, (Fort Collins: Colorado State University, 1981).

Weisbrod, B., "Collective-Consumption Services of Individualized-Consumption Goods," <u>Quarterly Journal of Economics</u>, 78 (August 1964), 471-77.

Wentz, Dennis A., <u>Effect of Mine Drainage on the Quality of Streams in Colorado, 1972-74</u>, Colorado Water Resources Circular No. 21, (Denver: Colorado Water Conservation Board, 1974).

Willeke, Gene E., <u>Effect of Water Pollution in San Francisco Bay</u>, Report No. EEP-29, Engineering, Economic Planning Program, (Palo Alto, Calif.: Stanford University, 1968).

Young, Robert A., S. L. Gray, et al., <u>Economic Value of Water: Concepts and Empirical Measurement</u>, Report to the National Water Commission by Colorado State University, Fort Collins, NTIS No. PB 210 356, (Springfield, Va.: U.S. Department of Commerce, March 1972).

Appendix A:
Concept of Option Demand

The origin of the concept of option demand may be traced to an article by Burton Weisbrod.[1] His formulation was in rebuttal of Milton Freidman's[2] advocacy of the extreme case of cutting down the redwoods in Sequoia National Park. Weisbrod set forth two conditions for the presence of option demand: (1) infrequency and uncertainty of demand for the commodity under consideration, and (2) prohibitively high cost in time or resources of renewing production of the commodity once it has been curtailed. Visits to Sequoia National Park are usually infrequent and uncertain. Should production of the magnificent forests be diverted from aesthetic enjoyment to lumbering, it would require centuries for the forest to become reestablished. The opportunity cost of lumbering would then be the aesthetic enjoyment foregone.

Weisbrod's analysis began with a simplified problem in which he assumed that a market existed for the collection of an admission fee from all users of the park. He assumed that the park was privately owned by a perfectly discriminating monopolist whose present value of total costs exceed present value of total revenues. All external economies were assumed away, the product was considered nonstorable, and the possibility of purchase before consumption was precluded. Given the foregoing propositions, if the private and social rates of discount were equal, then based solely on grounds of economic efficiency the park should be closed. Its productive resources should be reallocated to other uses.

Even so, Weisbrod contended that it may be unsound from society's standpoint to reallocate the park's resources. Given the presence of rational consumers who anticipate possibly visiting the park, but who are uncertain and in actuality may or may not make such a visit, they would be willing to pay a fee for an option which would guarantee their access to the park in the future. If a private market existed whereby this option value could be collected then it would influence the entrepreneur's decisions. However, without the option market, aggregate user fees would understate the total worth of the park to society. If in fact the park closes as a consequence of a lack of a practical way to collect the option value, the option demand of potential future users is unfulfilled.

Weisbrod emphasizes the fact that option value is significant for economic decision makers only when a decision to close the park is imminent. As long as the park remains open the provision of the option is a costless byproduct of current operation. It fulfills the conditions of a pure public good since all potential future users of the park can maintain the option without infringing on the consumption opportunities of others.

Following Weisbrod's introduction of option value, a debate ensued as to whether it was a totally new concept or merely "the unrecognized son of that old goat, consumer surplus."[3] Long attempted to show that option value was nothing more than "the expected consumer surplus from consuming the good at the terms specified in the option." He used Lerner's measure of consumer surplus as ". . . how much money a consumer would pay for the right to continue to buy at the current price something that he is now buying." He pointed out that the difference between his concept and Weisbrod's is that consumers under the latter definition may never use the option to purchase a commodity. The key to reconciling the two definitions, according to Long, is recognizing that the terms specified in the option will have a significant effect on option value.

Given a typical Marshallian individual demand curve, consumer surplus would normally be defined as the area under the demand curve and above price. According to Long, if price is a positive amount, option value will be of a smaller magnitude than if price is zero. If price should rise further to a level that the individual would never purchase the commodity because of its prohibitively high price, then no option fee would be paid to preserve future access to the good. Since option value is a fee paid for <u>future</u> access, Long concludes that option value is simply expected consumer surplus from consuming the commodity at the specified price.

Long contends that divisibility and homogeneity are the important concerns of option value and not frequency of use as Weisbrod indicated. In Long's view, option value attains significance only when discrete change in product must be made and no good substitute exists, rather than for marginal changes necessary for efficient resource allocation. Weisbrod's high cost condition then becomes unrelated to the problem. Long concludes that introducing option value into economic analysis would serve only to inflate measures of demand for public goods.

Lindsay[4] takes exception to Long's contention that "option value is exactly the expected consumer surplus from consuming the good at the terms specified in the option." He points out that Long ignores Weisbrod's initial assumption of uncertainty of consumption and implicitly substitutes certainty. Yet, as Lindsay explains, it would be nonsensical to purchase options for future consumption of goods which the consumer knows with certainty he will or will not purchase. But since uncertainty pervades the future, many consumers may wish to pay a premium to insure against the risk of not having the commodity available.

The debate continued with comments by Byerlee[5] and Cicchetti and Freeman.[6] Byerlee formulated the issue in mathematical terms which up to this point had been missing. Using a game theoretic framework, Byerlee established that under conditions of certainty of consumption the option demand was equivalent to consumer surplus. He further argued that where uncertainty of demand exists, option value as defined by Weisbrod can be shown to be greater than, equal to, or less than consumer surplus. He concludes that, as Long had suggested, including both consumer surplus and option demand would be double-counting. He supports a modification of Lerner's definition of consumer surplus to include "how much money a consumer would pay for the right to buy at the current price something that he is now buying or may buy in the future."[7]

Cicchetti and Freeman countered Byerlee by suggesting that along with uncertainty in demand, uncertainty in supply must also be considered. Possible deterioration of the natural amenities of the site must threaten the continuance of supply for option demand to become relevant. This was a condition specified for Weisbrod, which Byerlee neglected in his statement of the probability of option demand. Cicchetti and Freeman used probabilistic demand theory to illustrate that a risk-averse individual will be willing to pay a positive amount, to preserve his option of using a facility in the future, when there is a threat of an irreversible consequence to the natural environment. Cicchetti and Freeman then argue for the inclusion of a risk premium to be added to consumer surplus derived from recreational enjoyment of the site.

Schmalensee[8] points out that the other alternative must also be considered. If the natural environment is preserved, there is a risk associated with a very small future demand. Society may desire products which cause pollution of a site more than it desires recreation. The opportunity cost of preservation becomes very great. It is argued that the magnitude of each alternative risk determines whether option value is positive. Should the preservation alternative prove riskier then option value is negative. With no way of measuring the sign or magnitude of the associated risk premiums, Schmalensee regards expected consumer surplus as an adequate approximation of society's option values.

Two articles published at about the same time show that option value may exist without the side condition of risk aversion motivating an individual. If there is a prospect of better information forthcoming relating to alternative uses of an asset with irreplaceable characteristics, a situation arises where a positive option value may be generated. Arrow and Fisher[9] formulated a "quasi-option value" model developed in terms of aggregate benefits and costs of alternative environmental action. They questioned whether or not the existence of option value for the individual necessarily leads to a similar situation for society. They concluded that even in the aggregate society must take cognizance of the presence of option value. Henry[10] formulated a model based on individual willingness to pay for the assurance of selecting the preservation of an irreplaceable environmental asset facing an imminent irreversible commitment, until such time that sufficient information becomes available affecting the future option decision of selecting from among alternative competing uses.

Henry suggests that Schmalensee's "timeless world" assumption where an individual can make but one irreversible decision (one decision is as irreversible as another) reduces option value to "a risk premium in favor of 'irreplaceable assets'." Henry proposes a "sequential world" model where decisions must be made by the individual at appropriate intervals. In doing so, Henry adds to the conception of option value. The principle changes from paying to preserve the <u>option</u> of deciding later on the two alternative uses of the environment when conditions of certainty will exist. Henry's model serves as a theoretical basis for the empirical measurement of option demand in this study. A fuller presentation of the theory is provided below.

Consider a two time period version of Henry's model defined in accordance with the following symbols:

- N = the Nth individual
- U = N's utility function
- Y = N's income
- CS = N's consumer surplus generated from the use of the natural environment
- D = availability of the irreplaceable natural environment
  - D = d, the natural environment that is available
  - D = d*, the natural environment that has been appropriated for an alternative irreversible use and is unavailable
- OV = option value
- i = states of the world, i = 1,2
- j = time period, j = 1,2
- P_i = probability that state i will occur (where $\sum P_i = 1$ for i = 1,2)
- C^j = opportunity cost to retain the natural environment.

The model is predicated on the following assumptions: (1) the future is uncertain, (2) one use of the natural environment is more irreversible than the other, (3) a decision is imminent as to which of the two competing uses of the natural environment will be chosen, and (4) sequential decision making takes place based on improved information acquired through time. Let

$$U = \sum_{j=1}^{2} \sum_{i=1}^{2} P_i^j U_i^j (Y_i^j, D^j) \tag{A1}$$

be N's two period probability-weighted utility function. Assume an opportunity cost C^j must be paid to obtain $D^j = d$, that is, a cost is imposed in the form of foregone alternatives if the natural environment imposed is to remain available. C^1 and C^2 must be financed at instant 1 and instant 2, respectively, if $D^j = d$ is chosen. For simplicity of exposition C^j and Y^j are assumed to be known with certainty. The notation may then be simplified to $U_i^j(Y^j, d^*) = U_i^j(d^*)$ and in later equations, $U_i^j(Y^j - C^j, d) = U_i^j(d)$. Finally, assume that

$$\sum_{i=1}^{2} P_i^1 U_i^1(d) < \sum_{i=1}^{2} P_i^1 U_i^1(d^*). \tag{A2}$$

This assumption specifies that if only the first period is considered, N will choose d* so that the natural environment is not available. In this case, the cost C^1 of preserving the natural environment is greater than the associated benefits in period 1.

In the following case no new information is expected to become available between instant 1 and instant 2. A decision is made as if a "timeless world" existed. Consumer surplus for N can be defined as the equating factor in

$$\sum_{i=1}^{2} P_i^1 U_i^1(Y^1 - CS, d) + \sum_{i=1}^{2} P_i^2 U_i^2(d) = \sum_{j=1}^{2} \sum_{i=1}^{2} P_i^j U_i^j(d*). \quad (A3)$$

N will be willing to pay an amount CS at instant 1 to have d. Even after payment of CS, the individual will still receive the same expected utility as if the natural environment were not available.

N will choose the preserved natural environment as opposed to the development alternative if $CS > C^1$. No CS term need appear in the second period term of the preservation alternative, since no change in information occurs between the two periods. As long as C^1 is paid at instant 1, the natural environment will be available in all following periods because of the static situation. In this case CS is the present worth to N of the preserved natural environment for all time.

Now assume that new information enters between instant 1 and instant 2. Individual N will know with certainty at instant 2 which state of the world will obtain. Assuming that a sequential decision-making process takes place, the following question must be answered: How much will N be willing to pay at instant 1 to (1) enjoy the natural environment through period 1 and (2) to have the option of choosing under conditions of certainty at instant 2 whether or not to retain the natural environment?

The preceding question can be answered by referring to the following equation.

$$\sum_{i=1}^{2} P_i^1 U_i^1(Y^1 - CS^1 - OV, d) + \sum_{i=1}^{2} P_i^2 \max[U_i^2(d), U_i^2(d*)]$$

$$= \sum_{j=1}^{2} \sum_{i=1}^{2} P_i^j U_i^j(d*) \quad (A4)$$

The terms CS^1 and OV balance equation (A4). At instant 1 individual N will be willing to pay CS^1 to enjoy the natural environment during period 1. In addition, N is willing to pay an amount OV to choose, at instant 2, either the preserved environment or the irreversible alternative with full knowledge of which state of the world will obtain. In equation (A4) CS^1 results from the enjoyment of the preserved natural environment through period 1 only. The magnitude of OV in period 1 is a function of P_i^2, $U_i^2(d)$, and $U_i^2(d*)$ in period 2 as they exist at instant 1.

In considering the term $(\max[U_i^2(d), U_i^2(d^*)])$, four possible cases can occur:

1. $U_1^2(d) > U_1^2(d^*)$ and $U_2^2(d) < U_2^2(d^*)$
2. $U_1^2(d) < U_1^2(d^*)$ and $U_2^2(d) > U_2^2(d^*)$
3. $U_1^2(d) < U_1^2(d^*)$ and $U_2^2(d) < U_2^2(d^*)$
4. $U_1^2(d) > U_1^2(d^*)$ and $U_2^2(d) > U_2^2(d^*)$

For example, if situation (1) evolves, then

$$\sum_{i=1}^{2} [P_i^1 U_i^1 (Y^1 - CS^1 - OV, d)] + P_1^2 U_1^2(d) + P_2^2 U_2^2(d^*)$$

$$= \sum_{j=1}^{2} \sum_{i=1}^{2} P_i^j U_i^j(d^*). \tag{A5}$$

The inequality,

$$P_1^2 U_1^2(d) + P_2^2 U_2^2(d^*) > \sum_{i=1}^{2} P_i^2 U_i^2(d^*), \tag{A6}$$

exists because maximum values of $U_1^2(d)$ and $U_1^2(d^*)$ were chosen. Therefore, OV > zero and d will be chosen at instant 1 if $CS^1 + OV > C^1$. The magnitude of OV is determined precisely by the difference between the right- and left-hand expression of inequality (A6). In cases (2) and (3) OV will likewise be positive. Only in case (4) will option value equal zero. None of the four possible solutions will produce a negative option value.

Option value is irrelevant to the decision-making process as long as $CS^j > C^j$. The option to use the environment in the future has been preserved free of cost. Option value is a free byproduct as long as the user benefit of the preserved environment exceeds the opportunity costs of preservation. It is for this reason that inequality (A6) is required. This expression states that if the first period is considered by itself, development is preferred over preservation. Under this condition it is necessary to include explicit consideration of the second period in order to determine the proper course of action at the beginning of period 1. Henry extended the original analysis to include any number of sequential decision-making time periods. This empirical investigation, however, was limited to two time periods.

NOTES

1. Burton A. Weisbrod, "Collective-Consumption Services of Individualized-Consumption Goods," Quarterly Journal of Economics, 78

(August 1964), 471-77.

2. Milton Friedman, Capitalism and Freedom (Chicago: University of Chicago Press, 1962).

3. Millard F. Long, "Collective Consumption Services of Individual Goods: Comment," Quarterly Journal of Economics, 81 (May 1967), 351.

4. Cotton M. Lindsay, "Option Demand and Consumer's Surplus," Quarterly Journal of Economics, 83 (May 1969), 344-45.

5. D. R. Byerlee, "Option Demand and Consumer Surplus: Comment," Quarterly Journal of Economics, 85 (August 1971), 523-27.

6. Charles J. Cicchetti and A. M. Freeman, III, "Option Demand and Consumer Surplus: Further Comment," Quarterly Journal of Economics, 85 (August 1971), 529-39.

7. Long, op cit., 351.

8. Richard Schmalensee, "Option Demand and Consumer's Surplus: Valuing Price Changes Under Uncertainty," American Economic Review, 62 (December 1972), 814-24.

9. Kenneth J. Arrow and Anthony C. Fisher, "Environmental Preservation, Uncertainty, and Irreversibility," Quarterly Journal of Economics, 88 (May 1974), 312-19.

10. Claude Henry, "Option Values in the Economics of Irreplaceable Assets," The Review of Economic Studies: Symposium on the Economics of Exhaustible Resources, (1974), 89-104.

Appendix B: Questionnaire

FACSIMILE OF INTRODUCTORY LETTER

Department of Economics

Colorado State University
Fort Collins, Colorado
80523

    The Economics Department of Colorado State University is currently conducting a survey of the attitudes of Colorado residents toward the quality of water resources in the state.

    We are particularly interested in the values residents associate with water quality improvement. We want to find out how strongly you feel about water pollution. Since cleaning up bodies of water costs money, we would like to get a dollar estimate of how much clean water is worth to you.

    Your name has been randomly selected and a member of the Department will contact you shortly to request an interview at a time convenient to you.

    You have no legal obligation to cooperate in this survey. However, those who do may influence decisions about the quality of water in Colorado.

    We would sincerely appreciate your participation.

                                        Sincerely yours,

                                          Kenneth C. Nobe
                                          Chairman

KCN:mp

Department of Economics
Colorado State University
Fort Collins, CO 80523

WATER QUALITY OPINION SURVEY

1. How long have you lived in the Denver (Fort Collins) area? (years)
2. Where did you live before you moved to this area? City _____
    (1) Very large city or metro area (100,000+)?
    (2) Medium sized city (25,000-100,000)?
    (3) Small city (5,000-25,000)?
    (4) Rural area or town (nonfarm)?
    (5) Farm?
    (6) Have always lived in this area.
3. Why did you come here to live? (rank three most important)
    (1) A better job, higher income
    (2) Family
    (3) Cost of living
    (4) Climate
    (5) Recreation opportunities
    (6) Services
    (7) Health
    (8) Less pollution
    (9) Less congestion
    (10) School
    (11) Other
4. How many children do you have?
5. Which of the following best describes your family?
    (1) White
    (2) Oriental
    (3) American Indian
    (4) Black
    (5) Other
6. Is any of your family Spanish American or Mexican American?
7. Respondent's: (1) Sex (2) Age (3) Years of formal education (4) Employment (5) Employer (6) Head of household (0) No (1) Yes

    Coding:

Employment		Employer	
Professional	1	Small business	1
Business owner/mgr.	2	Large business	2
Skilled, foreman	3	Manufacturing	3
Salesman, buyer	4	Agriculture	4
Office worker	5	Petro/Chemicals	5
Unskilled	6	Mining	6
Housewife	7	Government	7
Retired	8	Unemployed	8
Student	9	Other	9
Other	0		

8. Which income category represents your total family income before taxes?
    (1) Under 6,000
    (2) 6,000-8,499
    (3) 8,500-10,999
    (4) 11,000-13,499
    (5) 13,500-15,499
    (6) 16,000-18,499
    (7) 18,500-20,999
    (8) 21,000 or more

9. Have you noticed any environmental problems in this area (e.g., air pollution, water pollution, land pollution (junk, dumps))? (0) No (1) Yes
10. Have any of these environmental problems affected your family or other people in this area (e.g., affected human health, damaged plants or livestock, reduced the enjoyment of life, or lowered property values)? (1) Nobody (2) Other people (3) Family (4) Other people and family
11. In general, how do you rate the waterways in the South Platte River Basin in terms of quality? (1) Poor (2) Fair (3) Good (4) Excellent
12. How would you classify lakes and streams in the South Platte River Basin as a source for your recreational enjoyment such as boating, swimming, fishing, or viewing? (1) Very important (2) Important (3) Somewhat important (4) Not important
13. If you were asked to distribute an increase in federal revenues, what percentage of 100 would you use to improve each of the following?

    (1) National defense
    (2) Highways
    (3) Education
    (4) Health services
    (5) Water quality
    (6) Air quality
    (7) Other

Coal development along with expanding mining operations may have significant effects on the quality of Colorado's water in the near future. As an aid in planning for the future I would like to find out how you feel about clean water for recreational activities. I have some questions which consider different ways of financing improvements of various levels of water quality. Let us consider three levels of water quality in a waterway such as the South Platte River Basin.

Suppose a sales tax was collected from the citizens of the South Platte River Basin for the purpose of financing water quality in this basin. All of the additional tax would be used for water quality improvements to enhance recreational enjoyment. Every basin resident would pay the tax. All bodies of water in the river basin would be cleaned up by 1983. Assume that this is the only way to finance water quality improvement.

14. Would you be willing to add ____ cents on the dollar to present sales taxes every year, if that resulted in an improvement from situation C to situation B?
15. Would you be willing to add ____ cents on the dollar to present sales taxes every year, if that resulted in an improvement from situation C to situation A?
16. (If 14 and 15 "zero" choose one.) Did you answer "zero" because:

    (1) You do not suffer any ill effects from water pollution and therefore see no reason to reduce it?
    (2) You believe taxes are already too high?
    (3) You believe it is unfair to expect people adversely affected to pay the costs of reducing water pollution?
    (4) Other

Now let's consider a different way of financing water quality improvement. Suppose an extra water bill charge was collected from citizens of the South Platte River Basin for the purpose of financing water quality in this basin. All of the additional charge would be used for water quality improvements to enhance recreational enjoyment. Every basin resident would pay the charge. All bodies of water in the river basin would be cleaned up by 1983. Now assume that this is the only way to finance water quality improvement.

17. Do you think it would be reasonable to add ____ to your water bill every month, if that resulted in an improvement from situation C to situation B?
18. Do you think it would be reasonable to add ____ to your water bill every month, if that resulted in an improvement from situation C to situation A?
19. (If 17 and 18 "zero" choose <u>one</u>.) Did you answer "zero" because:

    (1) You do not suffer any ill effects from water pollution and therefore see no reason to reduce it?
    (2) You believe your water bill is already too high?
    (3) You believe it is unfair to expect people adversely affected to pay the costs of reducing water pollution?
    (4) Other

20. If situation A was not achieved until the year 2000, how would this delay affect your payment of the (1) Sales tax from your original estimate of ____? (2) Water bill charge from your original estimate of ____?
21. (If all Colorado residents and tourists would pay the sales tax:)
    (If all Colorado water users would pay the water bill charge:) how would improving the quality of water to level A by 1983 <u>throughout Colorado</u> for recreational enjoyment affect your payment of the: (1) Sales tax from your original estimate of ____? (2) Water bill charge from your original estimate of ____?
22. How many days per year do you spend at water-based recreation such as fishing, boating, swimming, picnicking near streams, etc.,: (1) in the South Platte River Basin? (2) Anywhere in the United States?
23. How well do you like doing these activities?

    (1) Dislike very much   (4) Like
    (2) Dislike   (5) Like very much
    (3) Indifferent

24. What would you estimate are the chances in 100 that you will travel to lakes and streams in the South Platte River Basin in the next year, for water-based recreation if they are preserved at level A? (2) Do you anticipate any significant change in your chances for future years? (3) (If "yes") What change?
25. Given your chances of future use would you be willing to add ____ cents on the dollar to present sales taxes every year for water recreation opportunities at lakes and streams in the South Platte River Basin if they were preserved at level A?

Would it be reasonable to add ____ to your water bill every month for these opportunities?

In the near future, one of two alternatives is likely to occur in the South Platte River Basin. The <u>first alternative</u> is that a large expansion in mining development will soon take place, creating jobs and income for the region. As a consequence, however, many lakes and streams would become severely polluted. It is highly unlikely, as is shown in situation C, that these waterways could ever be returned to their natural condition. They could not be used for recreation. Growing demand could cause all other waterways in the area to be crowded with other recreationists.

The <u>second possible alternative</u> is to postpone any decision to expand mining activities which would irreversibly pollute these waterways. During this time, they would be preserved at level A for your recreational use. Furthermore, information would become available enabling you to make a decision with near certainty in the future, as to whether it is more beneficial to you to preserve the waterways at level A for your recreational use or to permit mining development. Of course, if the first alternative takes place, you could not make this future choice since the waterways would be irreversibly polluted.

26. Given your chances of future recreational use, would you be willing to add ____ cents on the dollar to present sales taxes every year to postpone mining development. This postponement would permit information to become available enabling you to make a decision with near certainty in the future as to which option (recreational use or mining development) would be most beneficial to you? Would it be reasonable to add ____ to your water bill every month for this postponement?
27. If it were certain you would not use the South Platte River Basin for water-based recreation, would you be willing to add ____ cents on the dollar to present sales taxes every year, just to know clean water exists at level A as a natural habitat for plants, fish, wildlife, etc.? Would it be reasonable to add ____ to your water bill every month for this knowledge?
28. If it were certain you would not use the South Platte River Basin for water-based recreation, would you be willing to add ____ cents on the dollar to present sales taxes every year to ensure that future generations will be able to enjoy clean water at level A? Would it be reasonable to add ____ to your water bill every month for this knowledge?
29. Who do you think should pay the costs of water quality preservation?

    (1) The people benefiting by it, i.e., the local residents and other recreationists.
    (2) The final consumer of the things produced by polluting industries.
    (3) The polluting firms.
    (4) Some combination of the above. (Which bears primary responsibility?)
    (5) The community as a whole.

Appendix C: Regression Analysis

SPECIFICATION OF THE REGRESSION MODELS

The hypothesis is that the dependent variable willingness to pay, Y, is a function of the specific socioeconomic characteristics of the survey participants. Independent socioeconomic variables X_1 through X_{30} were used in estimating the regression equations, although not all of the variables were found to be statistically significant.

Various algebraic forms including linear, hyperbolic, quadratic, and log functions were estimated. The linear model was chosen for the subsequent evaluations since it provided the best fit. The functions with insignificant constants were forced through the origin with no improvement in the percentage of explained variance or in the numbers of significant coefficients. Significant hyperbolic terms were checked for consistency through use of log transformations. Since consistent results were obtained the hyperbolic terms were retained in the final estimated equations.

Four models were considered. These models may be formulated as:

$$Y = a + Z_h + Z_i + Z_j + Z_k + B_1X_1 + B_2X_2 + \ldots, + B_nX_n + E \tag{C1}$$

where Y may be defined as:

MODEL I Resident household willingness to pay an increased sales tax for improved water quality to enhance recreation.
MODEL II Resident household willingness to pay an increased water bill for improved water quality to enhance recreation.
MODEL III Resident household willingness to pay an increased sales tax for option value of improved water quality.
MODEL IV Resident household willingness to pay an increased water bill for option value of improved water quality.

The regression parameters are:

a = joint reference category intercept value
Z_h = net effect of sex, $h = 1, 2$
Z_i = net effect of previous residence, $i = 1, \ldots, 4$
Z_j = net effect of occupation, $j = 1, \ldots, 4$
Z_k = net effect of employment, $k = 1, \ldots, 4$
B_n = regression coefficient for the continuous variables
E = stochastic disturbance variables

All four models are estimated for both the Denver metropolitan area and Fort Collins

Variables Used in Analysis of Relationship Between Willingness to Pay and Socioeconomic Characteristics

Y_1 Annual amount willing to increase sales tax for improved water quality
Y_2 Annual amount willing to increase water bill for improved water quality
Y_3 Annual amount willing to increase sales tax for option value
Y_4 Annual amount willing to increase water bill for option value
X_1 Gross annual family income
X_2 Annual South Platte River Basin water-based recreation days
X_3 Annual water-based recreation days, total
X_4 Age
X_5 Years of formal education
X_6 Female, 0 (dummy reference category)
X_7 1 if female, 0 otherwise
X_8 Population of previous residence -- under 5,000, 0 (dummy reference category)
X_9 1 if population of previous residence 5,000-25,000, 0 otherwise
X_{10} 1 if population of previous residence 25,000-100,000, 0 otherwise
X_{11} 1 if population of previous residence 100,000+, 0 otherwise
X_{12} Occupation -- all other including: foreman, salesman/buyer, office worker, unskilled, and student 0 (dummy reference category)
X_{13} 1 if occupation professional, business owner or manager, 0 otherwise
X_{14} 1 if occupation housewife, 0 otherwise
X_{15} 1 if occupation retired, 0 otherwise

X_{16} Employer -- all other including unemployed 0 (dummy reference category)
X_{17} 1 if employed in small business or agriculture, 0 otherwise
X_{18} 1 if employed in large business or manufacturing, 0 otherwise
X_{19} 1 if employed in government, 0 otherwise
X_{20} Number of children in family
X_{21} Years lived in present city
X_{22} 1/gross annual family income
X_{23} 1/age
X_{24} 1/years lived in present city
X_{25} 1/years of formal education
X_{26} Family income x age
X_{27} Family income squared
X_{28} Age squared
X_{29} Years of formal education squared
X_{30} Number of children in family squared

INTERPRETATION OF DUMMY VARIABLES

Four independent variables have distinct qualitative levels and are not subject to ordinary regression analysis as are the continuous quantitative independent variables. The four qualitative variables used in the stepwise regression analysis are: sex, size of previous residence, occupation, and employer. Variables X_6 through X_{19} serve as dummy variables for these qualitative categories. The number 1 is inserted in all but the base reference categories if the particular case fits a particular qualitative level, e.g., for a male respondent 1 becomes the value of X_7, 0 otherwise. Such a procedure using variables with N different qualitative categories will produce statistical results identical to an N-way analysis of variance, i.e., a test of significance for a dummy variable coefficient will yield the same conclusion as a test of whether the means of the population of the various levels are equal.

The intercept term in the regression model serves as a joint reference category. This category represents female, previous residence under 5,000, all other occupations, and all other employers. If one of the qualitative levels for a particular respondent deviates from this base category, the estimated dummy variable coefficient must be added to the base category to estimate the appropriate regression equation. For example, if one wished to estimate the willingness to pay for option value by a male, the estimated dummy variable coefficient for male must be added to the base reference category if it is statistically significant.

STATISTICAL RELIABILITY OF THE SOCIOECONOMIC REGRESSIONS

The least squares regression estimation procedure is based on the assumption that the disturbance terms are normally distributed with a zero mean and constant variance. There are a number of statistical problems associated with regression analysis of economic data which may violate this assumption resulting in biased and/or inconsistent estimators. Among these are multicollinearity, heteroscedasticity, misspecification of the model, nonrandom disturbance terms, and nonnormally distributed sample values. These problem areas will be briefly discussed in relation to the preceding analysis.

Multicollinearity arises when at least one of the independent variables is a linear combination of the others. This results in a situation of too few independent normal equations so that estimators cannot be derived for all of the coefficients. A classic symptom of multicollinearity occurs when a large coefficient of determination is produced while estimators of the coefficients are found to be insignificant. This finding might arise when high correlation exists between such variables as education or age and income. This was not the situation in the preceding analysis. Very low correlation was observed between the variables comprising the socioeconomic data generated from the survey. Only a very few pairs of variables had correlation coefficients exceeding 0.50. The symptom of insignificant coefficient estimators in conjunction with large R^2 values was not observed. Thus, there was no available evidence to suggest that multicollinearity resulted in any difficulty in estimating the regression coefficients.

The condition of nonconstant variance among the error terms is called heteroscedasticity. Since it is assumed in regression analysis that the variance of the error term is constant, this assumption is violated when heteroscedasticity occurs. The least squares estimation procedure produces an estimate of error term variance used in calculating the standard errors of the coefficients. These standard errors will become an average of differing variances of the error terms. Since the estimator does not produce the statistic with the smallest variance tests of hypothesis and confidence interval estimates derived by using the t-statistic will be suspect. The variance of the disturbance terms may be expected to increase as household income increases. Wealthy people generally have greater variability in their consumption patterns than do the poor. This situation was not apparent from a plot of the least squares residuals against willingness to pay estimates. Variance among the error terms appeared to be constant as willingness to pay estimates increased. There was no evidence to suggest the variance among the error terms was heteroscedastic.

Omission of an important independent variable may lead to a violation of the assumption of an expected zero mean for the error term. The low coefficients of determination of the various socioeconomic regressions might lead one to the conclusion that a significant variable not among those listed may have been omitted. One case in point is that there were no independent variables available which adequately served as an indicator of tastes and preferences of the participating residents regarding their enjoyment of the natural environment in general and water-based recreation activity in particular. Respondents were asked how well they like outdoor water-based recreation on a four-point scale from dislike very much to like very much. This scale did not provide sufficient detail to detect differences in the tastes of the respondents. Almost all residents indicated that they either "liked" or "liked water-based recreation very much." Likewise the number of water-based recreation activity days failed to indicate adequately the tastes of the residents for outdoor recreation. This was most likely the result of respondents who spent little time in these activities because of work or time restrictions but enjoyed water-based recreation activity very much. Other respondents reported spending a large number of days engaged in water-based recreation activity, not so much because they personally enjoyed the time as they wished to please other members of their families. Greater emphasis on the development of a refined question designed to reflect tastes and appreciation of the outdoor environment may mitigate this problem. Randall[1] reported a similar problem in a study of the benefits to abatement of air particulates in the Four Corners area of the Southwest.

The relatively low coefficients of determination may also be the result of omitting other significant variables associated with environmental concern. Our findings are less pessimistic than Eastman, Hoffer, and Randall who concluded that concern for the environment may tend to be randomly distributed among residents of the Four Corners area. The authors suggest that:[2]

> A tentative explanation of this randomness may be that attention has focused on the environment only recently and attitudes have not yet fully crystalized. With time, a more patterned relationship may emerge. Or it may be that aesthetic concerns are inherently less patterned than many other phenomena.

The fit of the regression equations in this study was somewhat higher than that of the equations estimated by Eastman, Hoffer, and Randall. A pattern, although somewhat weak, signified by the repeated occurrence of variables affecting willingness to pay was also more pronounced in this study. It may be that characteristics affecting environmental concern are crystalizing throughout the population but are still not fully identified. Interestingly, although it might be expected that a more homogeneous population would be found in a small university town, benefit estimates were more highly associated with socioeconomic profile characteristics of Denver residents. Perhaps attitudes toward the natural environment have crystalized more rapidly in the more densely populated Denver metropolitan area than in Fort Collins.

The willingness to pay bids generated from the survey were estimates of the amount of household income respondents would allocate to water quality improvement. They may have been subject to errors in measurement on the part of the respondents. This may have resulted in low R^2 values, as log linear, hyperbolic, and quadratic transformations did not improve the fit of the data.

There were no apparent trends in the overall plot of the residuals and plots of the residuals against the dependent and relevant independent variables. There was, however, an outlier problem associated with high bid estimates which could not be mitigated by use of any of the available independent variables. This problem may be related to the lack of a sufficient indicator of tastes or unpredictable behavior as discussed above.

Although plots of the estimated use and nonuse values appear to be normally distributed, there is a cluster of values at the zero point. Respondents were not permitted to associate negative values of recreational opportunities derived from water quality improvement. Hence the normal distribution curve of values becomes truncated instead of asymptotic. The 95 percent confidence intervals for the estimated coefficients of the independent variables are generally large. This problem reduces the predictive ability of the equations.

The statistical relationship between willingness of residents to pay for improved water quality and socioeconomic characteristics is estimated through use of variance estimators, based on sample responses to hypothetical situations. It is impossible to indicate what effect the introduction of a real, as opposed to hypothetical, payment situation would have on the willingness to pay variance estimators and hence the estimated statistical relationships.

NOTES

1. Letter from Alan Randall to Professor Richard G. Walsh, May 13, 1975, Department of Economics, Colorado State University, Fort Collins.

2. Clyde Eastman, Peggy Hoffer, and Alan Randall, A Socioeconomic Analysis of Environmental Concern: Case of the Four Corners Electric Power Complex, Bulletin 626, Agricultural Experiment Station (Las Cruces: New Mexico State University, September 1974), p. 22.

Table C-1. Stepwise Multiple Regression of Resident Households' Willingness to Pay Additional Sales Taxes to Improve Water Quality for Recreation Use, Denver, Colorado, 1976.

Dependent Variable: Amount of money per year respondent is willing to increase sales tax for improved water quality to enhance recreation. (Y_1)
Mean Bid: $50.18 Standard Error of $\hat{Y}_1$: 2.84
Sample: 85 residents
Statistical Summary: Fraction of explained variance (R^2): 0.4721
Regression degrees of freedom: 11   Residual degrees of freedom: 73
F ratio: 5.94   Significance: 0.0005

Independent Variable	Regression Coefficient	95% Confidence Interval	Standard Error	F to Remove	Significance
X_7 Male	26.55	11.60 – 41.50	7.50	12.53	0.001
X_{19} Employer-Government	28.44	10.69 – 46.19	8.91	10.20	0.002
X_{17} Employer-Small Business	25.08	7.06 – 43.10	9.04	7.70	0.007
X_{15} Occupation-Retired	-30.03	-52.00 – -8.07	11.02	7.43	0.008
X_{13} Occupation-Professional/ Business Owner-Manager	-21.67	-39.05 – -4.29	8.72	6.18	0.015
X_{26} Family Occupation x Age	0.00001	0.0000026 – 0.000029	0.000006	5.74	0.019
X_{14} Occupation-Housewife	20.91	1.98 – 39.84	9.50	4.85	0.031
X_{11} Previous Residence (100,000+)	-13.59	-28.18 – 1.11	7.37	3.40	0.069
X_2 South Platte Recreation	0.22	-0.16 – 0.60	0.19	1.35	0.249
X_4 Age	0.29	-0.25 – 0.82	0.27	1.13	0.292
X_{10} Previous Residence (25,000-100,000)	-8.27	-24.71 – 8.18	8.25	1.00	0.320
(Constant)	17.24	-8.67 – 43.15	13.00	1.76	0.189

Table C-2. Stepwise Multiple Regression of Resident Households' Willingness to Pay Higher Water Service Prices to Improve Water Quality for Recreation Use, Denver, Colorado, 1976.

Dependent Variable: Amount of money per year respondent is willing to increase water bill for improved water quality to enhance recreation. (Y_2)

Mean Bid: $15.84 Standard error of $\hat{Y}_2$: 1.20
Sample: 82 residents
Statistical Summary: Fraction of explained variance (R^2): 0.3518
Regression degrees of freedom: 6   Residual degrees of freedom: 75
F ratio: 6.78   Significance: 0.0005

Independent Variable	Regression Coefficient	95% Confidence Interval	Standard Error	F to Remove	Significance
X_{24} 1/Years Lived in City	50.34	25.28 – 75.39	12.58	16.01	<0.0005
X_7 Male	6.92	1.43 – 12.41	2.76	6.30	0.014
X_{19} Employer-Government	7.18	0.53 – 13.82	3.34	4.62	0.035
X_{14} Occupation-Housewife	6.21	-0.65 – 13.08	3.45	3.25	0.075
X_4 Age	-0.13	-0.28 – 0.03	0.08	2.63	0.109
X_1 1/Family Income	18,360	-9,879 – 46,600	14,176	1.68	0.199
(Constant)	8.79	-0.43 – 18.01	4.63	3.60	0.062

Table C-3. Stepwise Multiple Regression of Resident Households' Willingness to Pay Higher Water Service Prices to Improve Water Quality for Recreation Use, Fort Collins, Colorado, 1976.

Dependent Variable: Amount of money per year respondent is willing to increase water bill for improved water quality to enhance recreation. (Y_2)

Mean Bid: $25.92
Sample: 78 residents
Statistical Summary: Fraction of explained variance (R^2): 0.1949   Standard error of $\hat{Y}_2$: 3.23
Regression degrees of freedom: 4   Residual degrees of freedom: 73
F ratio: 4.42   Significance: 0.003

Independent Variable	Regression Coefficient	95% Confidence Interval	Standard Error	F to Remove	Significance
X_5 Education	8.44	2.95 - 13.93	2.75	9.38	0.003
X_{26} Family Income x Age	0.00002	-0.000005 - 0.00005	0.00001	2.72	0.103
X_{11} Previous Residence (100,000+)	16.23	-15.51 - 47.97	15.93	1.04	0.311
X_3 Annual Water Recreation	-0.26	-0.730 - 0.24	0.24	1.03	0.313
(Constant)	-11.45	-195.28 - -27.61	42.07	7.02	0.010

153

Table C-4. Stepwise Multiple Regression of Resident Households' Willingness to Pay Additional Sales Taxes for Option Value from Preserved Water Quality, Denver, Colorado, 1976.

Dependent Variable: Amount of money per year respondent is willing to increase tax for option value. (Y_3)

Mean Bid: $18.31
Sample: 88 residents
Statistical Summary: Fraction of explained variance (R^2): 0.2857
Regression degrees of freedom: 10
F ratio: 3.08
Standard error of $\hat{Y}_3$: 2.28
Residual degrees of freedom: 77
Significance: 0.002

Independent Variable	Regression Coefficient	95% Confidence Interval	Standard Error	F to Remove	Significance
X_4 Age	-0.66	-1.02 – -0.30	0.18	13.34	<0.0005
X_1 Family Income	0.00068	0.00019 – 0.00117	0.00024	7.66	0.007
X_9 Previous Residence (5,000-25,000)	17.70	4.36 – 31.04	6.69	6.99	0.010
X_{13} Occupation-Professional/Business Owner-Manager	-14.07	-26.94 – -1.20	6.47	4.74	0.033
X_3 Annual Water Recreation	-0.17	-0.37 – 0.03	0.10	2.71	0.104
X_{18} Employer-Large Business/Manufacturing	-10.66	-25.30 – 3.99	7.36	2.10	0.152
X_{20} Number of Children	2.37	-0.67 – 5.40	1.52	2.41	0.124
X_7 Male	8.05	-3.64 – 19.75	5.87	1.88	0.174
X_{24} 1/Years Lived in City	-26.67	-76.60 – 23.25	25.07	1.13	0.291
X_{19} Employer-Government	6.94	-6.28 – 20.16	6.64	1.09	0.299
(Constant)	35.93	16.18 – 55.67	9.91	13.13	0.001

Table C-5. Stepwise Multiple Regression of Resident Households' Willingness to Pay Additional Sales Taxes for Option Value from Preserved Water Quality, Fort Collins, Colorado, 1976.

Dependent Variable: Amount of money per year respondent is willing to increase sales tax for option value. ($\hat{Y}_3$)

Mean Bid: $34.05
Sample: 89 residents
Standard Error of $\hat{Y}_3$: 6.55

Statistical Summary: Fraction of explained variance (R^2): 0.1678
Regression degrees of freedom: 4 Residual degrees of freedom: 84
F ratio: 4.23 Significance: 0.004

Independent Variable	Regression Coefficient	95% Confidence Interval	Standard Error	F to Remove	Significance
X_7 Male	38.04	10.48 - 65.61	13.86	7.53	0.007
X_1 Family Income	0.00153	0.00009 - 0.00298	0.00072	4.49	0.037
X_{20} Number of Children	-10.45	-20.40 - -0.50	5.00	4.36	0.040
X_{17} Employer-Small Business, Agriculture	-25.49	-55.82 - 4.84	15.25	2.79	0.098
(Constant)	10.98	-21.90 - 43.87	16.54	0.44	0.508

Table C-6. Stepwise Multiple Regression of Resident Households' Willingness to Pay Higher Water Service Prices for Option Value from Preserved Water Quality, Denver, 1976.

Dependent Variable: Amount of money per year respondent is willing to increase water bill for option value. (Y_4)

Mean Bid: $6.04
Sample: 83 residents
Standard Error of $\hat{Y}_4$: 0.61

Statistical Summary:
Fraction of explained variance (R^2): 0.3809
Regression degrees of freedom: 9
Residual degrees of freedom: 73
F ratio: 4.99
Significance: 0.0005

Independent Variable	Regression Coefficient	95% Confidence Interval	Standard Error	F to Remove	Significance
X_4 Age	-0.15	-0.25 – -0.05	0.04	10.23	0.002
X_9 Previous Residence (5,000-25,000)	5.02	1.55 – 8.49	1.74	8.35	0.005
X_7 Male	3.65	0.68 – 6.61	1.49	6.01	0.017
X_{19} Employer-Government	3.91	0.39 – 7.44	1.77	4.92	0.030
X_{22} 1/Family Income	15,176	454 – 29,898	7,386	4.22	0.043
X_{18} Employer-Large Business/Manufacturing	-3.39	-7.25 – 4.78	1.94	3.05	0.085
X_{25} 1/Education	-80.58	-180.80 – 19.65	50.29	2.57	0.113
X_{13} Occupation-Professional/Business Owner-Manager	-2.35	-5.95 – 1.23	1.80	1.70	0.197
X_{15} Occupation-Retired	-2.34	-6.78 – 2.09	2.23	1.11	0.296
(Constant)	15.65	7.45 – 23.85	4.11	14.47	0.000

Table C-7. Stepwise Multiple Regression of Resident Households' Willingness to Pay Higher Water Service Prices for Option Value from Preserved Water Quality, Fort Collins, Colorado, 1976.

Dependent Variable: Amount of money per year respondent is willing to increase water bill for option value. (Y_4)
Mean Bid: $12.00   Standard Error of $\hat{Y}_4$: 2.76
Sample: 78 residents
Statistical Summary: Fraction of explained variance (R^2): 0.2390
Regression degrees of freedom: 4   Residual degrees of freedom: 73
F ratio: 5.73   Significance: 0.0005

Independent Variable	Regression Coefficient	95% Confidence Interval	Standard Error	F to Remove	Significance
X_5 Education	2.71	0.40 – 5.03	1.16	5.44	0.022
X_{22} 1/Family Income	55,377.66	-13,439 – 124,195	34,529	2.57	0.113
X_7 Male	8.56	-3.81 – 20.93	6.21	1.90	0.172
X_{20} Number of Children	-2.65	-7.20 – 1.90	2.28	1.35	0.250
(Constant)	-34.80	-70.77 – 1.17	18.05	3.71	0.058

Index

Adams, Gerard F., 17
Aesthetic benefits, 12, 25, 27, 29, 70, 78, 110, 116, 131, 149
Alaska, study of wilderness benefits in, 29, 65
Arizona, 70
  See also Four Corners Area
Arrow, Kenneth J., 26, 33(n. 45), 133

Battelle Memorial Institute, 24
Bell, Frederick W., 24
Benefits of pollution control,
  See Bequest value; Existence value; Option value; Preservation value; Recreation value; South Platte River Basin Study
Bequest value, 3, 5, 24-25, 115-118
  as a public good, 27, 28
  defined, 5, 25, 27, 117
  measurement of, 6, 12-13, 61-62, 63, 70-71, 74, 75, 77, 115, 116, 117-118
  See also Nonuse benefits; Preservation value
Bias,
  in response, 13, 71
    See also Free rider problem
  in sampling, 45-49
  statistical, 148-150
Bids,
  iterative, 6, 59-62, 115, 116
  outlier problem, 150
  zero, 50, 53-54

Bohm, Peter, 13, 78
Boston, study of value of water quality to residents of, 15, 75
Bouwes, Nicolaas W., 20, 69
Bradford, David F., 3
British Columbia, Canada
  See Fraser River
Brookshire, David S., 29, 65
Brown, Gardner S., 16
Byerlee, D. R., 132, 133

Canterberry, Ray E., 24
Charles River Basin, study of benefits of improving water quality in, 15-16, 74-75
Cicchetti, Charles J., 132, 133
Clawson, Marion, 30(n. 4)
Clean Water Act of 1977, 1
Colorado,
  mining in, 37, 39-40, 51
  reasons for moving to, 104
  recreation in, outdoor, 38, 113, 115
  study of wilderness benefits in, 29-30, 65
  value of improved water quality in, 84, 118-119
  water quality in, 2, 38-39, 42, 51
    See also Mining, effects on water quality of
  Water Quality Control Act of 1973, 38
  Water Quality Standards of 1974, 38, 43
  See also Denver; Fort Collins;

159

Rocky Mountain National Park;
South Platte River Basin
Consumer surplus,
as a measure of recreation benefits, 5, 9-11, 28
definition, 62(n. 4), 132, 133
nonequivalence to option value, 26, 132-136
Contingent value approach, 11-13, 15-16
methodology of, 11, 12-13, 45, 50-62, 115
See also South Platte River Basin Study, methodology of
strategic behavior by respondents to, potential, 13
See also Free rider problem
studies of willingness to participate using, 16
studies of willingness to pay using, 15-16, 29, 65, 69, 74-75, 116, 121
See also South Platte River Basin Study
Costs of pollution control, 3-4, 87, 90, 119

Data collection procedures, 6, 45-62, 115-116
contacting potential respondents, 49
interviewing, 50-62
selecting sample, 45
See also Bias; Bids; Contingent value approach, methodology; Payment vehicles
Davidson, Paul, 17
Davis, Robert K., 13, 69
Delaware River, study of benefits of improving water quality in, 17-18
Denver, 35, 37, 45-46, 49, 54, 64, 71, 74, 77-79, 84, 87, 90-91, 96-113, 115, 119-121, 149
Discount rate, 77, 118
Ditton, Robert, 16
Dummy variables, 147

Eastman, Clyde, 149
Environmental Protection Agency (EPA), 1, 2, 22, 84
Ericson, Ray K., 15
Eubanks, Larry S., 29, 65

Existence value (existence benefits), 3, 5, 24-25, 26-27, 115-118
as a public good, 27
defined, 5, 25, 26
measurement of, 12-13, 29, 61, 63, 70-71, 74, 75, 77, 115, 116, 117-118
See also Nonuse benefits; Preservation value
Extrapolation of benefits,
over time, 6, 63-64, 65, 69, 71, 75, 115
to national estimates, 22-24, 70, 71-74, 92(n. 9)
See also Present value; Weighting of samples

Federal water acts,
See Clean Water Act of 1977; Water Pollution Control Act of 1972
Feenberg, Daniel, 23
Fisher, Anthony C., 26, 27, 33(n. 45), 133
Fort Collins, 37, 45-46, 50, 54, 71, 74, 77, 78, 84, 87, 90-91, 96-113, 115, 119-121, 149
Four Corners Area, studies in value of aesthetic damages in, 13, 70, 149
Fraser River, study of benefits from, 28, 74
Free rider problem, 13, 53, 78, 118
See also Public good
Freeman, A. Myrick III, 13, 23, 24, 29, 69-70, 71, 74, 92(n. 9), 132, 133
Friedman, Milton, 25, 131

Gibbs, K. C., 18
Gillman, Richard A., 29-30, 65
Glen Canyon National Recreation Area,
See Powell, Lake
Goodale, Thomas, 16
Gramlich, Fred W., 15, 75
Gravity model, 22
Green Bay (Lake Michigan), study of water pollution impact on recreation participation in, 16

Heavy metal content, as index of water quality, 39, 50-51, 79, 116
Heintz, H. T., 22
Henry, Claude, 5, 26, 133, 134
Hershaft, A., 22
Hoffer, Peggy, 149
Horak, G. C., 22
Hotelling, Harold, 30(n. 4)

Idaho, study of benefits of improving water quality in, 21-22
Imminence of change and preservation values,
See Public good
Inverse demand function, for water quality, 12
Irreversibility, 5, 39, 115
See also Option value, definition of

Klamath Lake, Upper, study of benefits of improving water quality in, 18-19
Knetsch, Jack L., 23, 30(n. 4), 69
Krutilla, John V., 3, 5, 25, 26-27, 30

Lerner, Lionel J., 132, 133
Lindsay, C. M., 132
Liu, Ben-chieh, 21
Long, M. F., 132, 133
Loomis, John B., 29-30, 65
Low, Christopher R., 29, 65

Massachusetts,
See Charles River Basin; Merrimack River Basin
Mathews, B. S., 16
McKean, John R., 15
Merrimack River Basin, study of benefits of improving water quality in, 15, 69
Meyer, Phillip A., 28, 49
Michigan, Lake,
See Green Bay
Mills, Edwin, 23
Mining,
effects on water quality of, 40-42, 51, 53, 116
in Colorado, 37, 39-40, 51
in the South Platte River Basin, 37, 39, 51, 60, 116
Minneapolis-St. Paul, study of benefits of improving water quality in, 21
Morgan, Robert E., 41

Nashua River, study of benefits of improving water quality in, 19
National Commission on Water Quality, 3, 23, 24
National Planning Association, 24
New Hampshire,
See Merrimack River Basin; Nashua River
New Mexico, studies of value of aesthetic damages in, 13, 70, 149
Nonuse benefits, 12, 27-29, 71, 74
See also Preservation values

Option value (option benefits), 3, 5, 16, 24-26, 44(n. 4), 64-65, 116-118, 131-136
as a public good, 26, 132
definition of, 5, 25, 26, 64, 117, 131-134
measurement of, 6, 12-13, 26, 29, 60-61, 63, 64-65, 70, 74, 75, 77, 92(n. 9), 116-117, 134-136
See also South Platte River Basin Study
nonequivalence to consumer surplus of, 26, 132-136
of big game hunting, 29, 65
of wilderness, 29, 65
socioeconomic variables tested for significance in explaining, 93-113
See also Socioeconomic variables
theory of, 25-26, 131-136
See also Nonuse benefits; Preservation value
Oregon, study of benefits of improving water quality in, 21
See also Yaquina Bay; Klamath Lake, Upper
Oster, Sharon, 15, 69

Participation in recreation, 113
by sex, 97, 120

Participation in recreation,
  models of, 17-18, 22, 24
  willingness for, estimated by
    contingent value approach, 16
  See also Colorado, recreation
    in, outdoor; South Platte
    River Basin, recreation in,
    water-based
Payment vehicles, 15, 51-54, 77-
  78, 118
Perception of water quality,
  See Water quality, indexes of
Pike Lake, study of benefits of
  improving water quality in,
  20-21, 69
Pollution,
  air, 13, 70, 90, 149
  water, 1-2, 35-42
    See also Benefits of pollution
      control; Costs of pollution
      control; Water quality
Powell, Lake, study of value of
  air pollution damage at, 13,
  70
Present value, 75-77, 117-118
Preservation values (preservation
  benefits), 3, 5, 12, 16, 24-
  30, 70-71, 74, 115, 116-117
  measurement of, 11, 12-13, 15-
    16, 28-30, 63, 70-71, 74-75,
    77, 117
  of salmon, 28
  of wilderness, 29
  See also Bequest value; Exis-
    tence value; Option value
Public good (collective good), 9,
  26, 27, 28, 132
  See also Free rider problem

Randall, Alan, 13, 29, 65, 149
Recreation site attractiveness
  model, 22
Recreation value (recreation benefits), 9-24
  as consumer surplus, 5, 9-11
    See also Recreation value,
      contingent value method of
      measurement, travel cost
      method of measurement, unit
      day method of measurement
  contingent value method of
    measurement, 11, 12-13, 15-
    16, 29, 59-60, 69, 74-75, 115,
    116
    See also Contingent value
      approach
  estimates of, 6-7, 13, 14-24,
    28, 29, 30, 63-66, 69-70, 74-
    75, 77, 78, 90-91, 92(n. 9),
    116, 117-118
  household production function
    model method of measurement,
    24
  of boating, 16, 17, 17-18, 18,
    21-22, 22, 24
  of camping, 21-22
  of fishing, 14, 16, 17, 18, 21-
    22, 22, 24, 28
  of hunting, 14, 22, 29
  of swimming, 16, 17, 18, 21, 21-
    22, 22
  of water skiing, 18
  of wilderness, 14, 30
  sensitivity analysis of, 17
  socioeconomic variables tested
    for significance in explain-
    ing, 93-113
    See also Socioeconomic vari-
      ables
  travel cost method of measure-
    ment, 11, 11-12, 18-23, 24,
    30, 30(n. 4), 69
  unit day method of measurement,
    11, 17, 24
  See also Participation in rec-
    reation, models of
Regression analysis,
  See Contingent value approach;
    Travel cost approach; Parti-
    cipation in recreation, models
    of; Socioeconomic variables
Reiling, S. D., 18
Risk and option value, 132-134
Rocky Mountain National Park, 37,
  65, 69
  study of recreation benefits of
    water quality in, 15, 20, 65,
    69
Russell, Clifford S., 19

Schmalensee, R., 133
Schneider, Robert, 20, 69
Seneca, Joseph, 17
Sensitivity analysis, 17
Sequoia National Park
  See Friedman, Milton; Option

value; Weisbrod, Burton A.
"Sequential world" model, 5, 134-136
Socioeconomic variables,
  age, 96, 107, 120
  education, 101, 120
  employer, 97-98, 120
  in contingent value method studies, 12, 15
    See also South Platte River Basin Study, socioeconomic variables tested in
  in participation models, 17-18, 21-22, 24
  in travel cost method studies, 11, 19-20, 22
  income, household, 93, 96-97, 119
  move to Colorado, reason for, 104, 121
  occupation, 98, 120
  participation in water-based recreation, 91, 110, 113, 119
    See also South Platte River Basin, recreation in, water-based
  permanence of residence, 104, 107, 120
  previous residence, size of city of, 101, 121
  sex, 97, 120
  size of household, 107, 110, 120
South Platte River Basin, 35-39
  description of, physical and economic, 35-38, 39-40, 64, 115
  economic benefits of improved water quality (public perception of option and preservation values) in,
    See Rocky Mountain National Park, study of recreation benefits of water quality in; South Platte River Basin Study
  recreation in, water-based, 35, 38, 84, 90, 91, 110, 113, 115, 116
  water quality in, 35, 37, 38-39, 42, 51, 53, 79, 84, 90, 115-116
  See also Denver; Fort Collins; Mining, in South Platte River Basin; Rocky Mountain National Park
South Platte River Basin Study,
  effect of delaying water quality improvement on results of, 79-80, 118
  estimates of benefits of improved water quality in other river basins to residents of South Platte River Basin, 75, 84, 118-119
  estimates of benefits of improved water quality in South Platte River Basin to residents, 6-7, 64-79, 116-118
  methodology of, 6, 45-64, 71, 74, 75, 77, 90, 115-117, 145-150
    See also Data collection procedures; Discount rate; Extrapolation of benefits
  nonsignificance of inter-city benefit difference in, 90-91, 119
  objectives of, 5, 115
  responsibility for costs of water quality improvement, in opinion of respondents to, 87, 119
  socioeconomic variables tested in, 7, 90-91, 93-113, 119-121, 145-150
    See also Socioeconomic variables
  study area, selection criteria for, 35-36
  theoretical basis of,
    See Contingent value approach; "Sequential world" model
  water quality index used in, 39, 50-51, 78-79, 116
  See also Rocky Mountain National Park; South Platte River Basin
Stevens, Joe B., 18
Stoevener, H. H., 18
Strategic behavior, 13
  See also Free rider problem
Sutherland, Ronald J., 21, 22
Substitution, 23
  See also Contingent value approach; Travel cost approach; Unit day approach

Texas, study of demand for recreation at reservoirs in, 19-20
Tihansky, Dennis P., 17
Travel cost approach,
　See Recreation value, travel cost method of measurement

Uncertainty,
　See Option value, theory of
Unger, Samuel G., 23
Unit day approach,
　See Recreation value, unit day method of measurement
United States government agencies,
　See Environmental Protection Agency (EPA); National Commission on Water Quality; U.S. Water Resources Council
U.S. Water Resources Council, procedures for valuing recreation and environmental quality recommended by, 6, 11, 12, 13-14, 23, 24, 45, 115
Update,
　See Extrapolation of benefits, over time
Utah,
　See Powell, Lake

Walsh, Richard G., 15, 20, 22, 23, 29-30, 65
Washington (state), studies of recreation value changes due to water quality changes in, 16, 21
Water Pollution Control Act of 1972, 1, 24, 60
Water quality,
　benefits of improving,
　　See Bequest value; Existence value; Option value; Preservation value; Recreation value; South Platte River Basin Study
　decreasing marginal value of, 78-79, 118
　goals,
　　See Clean Water Act; Water Pollution Control Act of 1972
　indexes of, 2, 6, 15, 16, 20-21, 39, 75
　See also Heavy metal content standards for, 2, 38, 42, 43
Water Resources Council,
　See U.S. Water Resources Council
Weighting of samples, 64
Weisbrod, Burton A., 3, 25-26, 30, 131, 132, 133
Wentz, Dennis A., 40-41, 42
Wisconsin,
　See Green Bay (Lake Michigan); Pike Lake
Wyoming, study of wildlife benefits in, 29, 65

Yaquina Bay, Oregon, study of benefit loss to water pollution in, 18
Young, Robert A., 15

HD
1695
.S63
G74
1982